PROGRESS IN MUTATION RESEARCH

VOLUME 2

PROGRESS IN MUTATION RESEARCH

Volume 1

EVALUATION OF SHORT-TERM TESTS FOR CARCINOGENS
Report of the International Collaborative Program
edited by Frederick J. de Serres and John Ashby

Volume 2

PROGRESS IN ENVIRONMENTAL MUTAGENESIS AND CARCINOGENESIS
Proceedings of the 10th Annual Meeting of the European Environmental
Mutagen Society (EEMS), Athens 1980
edited by A. Kappas

Volume 3

CHEMICAL MUTAGENESIS, HUMAN POPULATION MONITORING AND
GENETIC RISK ASSESSMENT
edited by K.C. Bora

Volume 4

DNA REPAIR, CHROMOSOME ALTERATIONS AND CHROMATIN
STRUCTURE
edited by A.T. Natarajan

PROGRESS IN MUTATION RESEARCH

VOLUME 2

PROGRESS IN ENVIRONMENTAL MUTAGENESIS AND CARCINOGENESIS

Proceedings of the 10th Annual Meeting of the European Environmental Mutagen Society (EEMS)
Athens (Greece), 14–19 September 1980

Under the auspices of the Ministry of Culture and Science of Greece and the Greek Atomic Energy Commission

edited by

A. Kappas

Biology Department, Nuclear Research Centre "Democritus", Athens (Greece)

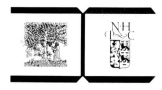

1981
Elsevier/North-Holland Biomedical Press
Amsterdam · Oxford · New York

© 1981 Elsevier/North-Holland Biomedical Press

All rights reserved. No part of this publication may be reproduced, stored in a retrieval system, or transmitted, in any form or by any means, electronic, mechanical, photocopying, recording or otherwise, without the prior permission of the copyright owner.

ISBN Vol. 0-444-80334-3
ISBN Ser. 0-444-00570-6

Published by:
Elsevier/North-Holland Biomedical Press
1 Molenwerf, 1014 AG, P.O. Box 211
Amsterdam, The Netherlands

Sole distrubutors for the USA and Canada:
Elsevier North Holland Inc.
52 Vanderbilt Avenue
New York, N.Y. 10017

Printed in The Netherlands

PREFACE

The European Environmental Mutagen Society (EEMS) was founded in 1969 and its 10th annual meeting was held in Athens, Greece from 14th to 19th of September 1980.

Scientists actively involved in the field of genetic toxicology from many European countries, the United States, Japan and others were present. Their scientific contribution was shown in the form of posters, oral presentations and invited lectures. Obviously the continuous and active discussions developed during the six days of the meeting was the best result of this meeting.

The invited lectures presented during the meeting dealt mainly with the aspect of mutagenic potential of food and pesticides, the mutagenic and carcinogenic evaluation of environmental pollutants and several methodological and basic problems common to the different topics of this field of research.

The text of these lectures is given in this book and they clearly represent the present achievements in the field of environmental mutagenesis and carcinogenesis.

The abstracts of posters and oral presentations will appear in *Mutation Research* (Vol. 85, No. 4, August 1981, pp. 215–307).

The organization of the meeting was excellent and all participants could appreciate the values of the cultural tradition of Greece which has seen the birth of modern science and philosophy.

N. LOPRIENO
President of EEMS

SCIENTIFIC PROGRAMME COMMITTEE

N.P. Bochkov, B.A. Bridges, U.H. Ehling, A. Kappas, P. Oftedal, M. Sorsa, L.G. Zetterberg

ORGANIZING COMMITTEE

President: A. Kappas
Officers: M. Pelecanos, C. Sekeris, C. Kastritsis, E. Sideris
Members: K. Athanasiou, N. Demopoulos

ACKNOWLEDGMENT

Financial support for the Meeting was provided by the

    Ministry of Culture and Science of Greece
    Greek Atomic Energy Commission
    National Tourist Organization of Greece
    Olympic Airways
    Commission of the European Communities
    L'Oreal, France
    Anelor, Greece
    Imperial Chemical Industries Ltd., United Kingdom
    Sandoz Ltd., Switzerland
    Petrola Hellas S.A., Greece
    Hoechst Hellas, Greece
    Wellcome Foundation Ltd., Greece
    Ciba-Geigy Hellas, Greece
    Zootechniki S.A., Greece
    Shell Chemicals Hellas Ltd., Greece

CONTENTS

Preface v

Scientific Programme Committee; Organizing Committee; Acknowledgement vi

Mutagens in Food and Gut Contents

Faecal mutagens: Their discovery and possible relevance to the aetiology of large-bowel cancer in man
    by S. Venitt (Chalfont St. Giles) 3
Mutagenic and colicin-inducing activity of some antioxidants
    by Renana Ben-Gurion (Tel-Aviv) 11
Acetaldehyde not ethanol is mutagenic
    by G. Obe (Berlin) 19
Penicillium roqueforti toxin
    by S. Moreau (Villeneuve d'Ascq) 25
Biological activity of the aflatoxins
    by R. Colin Garner (York) 33
The role of bacterial metabolism in the gut in relation to large bowel cancer
    by M.H. Thompson and M.J. Hill (London and Salisbury) 41
Mutagen—carcinogens in amino acid and protein pyrolysates and in cooked food
    by T. Matsushima and T. Sugimura (Tokyo) 49

Mutagenic Pesticides

The genotoxicity of benomyl
    by A. Kappas (Athens) 59
Does dichlorvos constitute a genotoxic hazard?
    by Claes Ramel (Stockholm) 69
The mutagenicity of the fungicide thiram
    by M. Zdzienicka, M. Zielenska, M. Hryniewicz, M. Trojanowska, M. Zalejska and T. Szymczyk (Warsaw) 79
Comparison of the mutagenic activity of pesticides in vitro in various short-term assays
    by A. Carere and G. Morpurgo (Rome) 87
Case-control studies: Soft-tissue sarcomas and malignant lymphomas and exposure to phenoxy acids or chlorophenols
    by L. Hardell (Umeå) 105

Mutagenic and Carcinogenic Pollutants

Cigarette smoke induced DNA damage in man
  by H.J. Evans (Edinburgh) 111
How relevant are high doses in mutagenicity and carcinogenicity studies in animals?
  by H. Greim, U. Andrae, W. Göggelmann, S. Hesse, L.R. Schwarz and K.H. Summer (Neuherberg). 129

Fundamental Microbiological Aspects of the Ames Test
Panel Discussion

Introduction
  by J.P. Seiler (Wädenswil) 151
Influence of composition and treatment of the growth media on the yield of mutant colonies
  by J.P. Seiler (Wädenswil) 155
The effect of spontaneous mutation on the sensitivity of the Ames test
  by M.H.L. Green (Brighton) 159
Test cell problems in the Ames test
  by A. Grafe (Mannheim) 167
Culture conditions and influence of the number of bacteria on the number of revertants
  by W. Göggelmann (Neuherberg). 173
Statistical problems in the Ames test
  by J. Vollmar (Mannheim) 179
Basis of evaluation of an Ames test
  by I.E. Mattern (Rijswijk). 187
Summary and discussion
  by J.P. Seiler (Wädenswil) 191

Concluding Remarks

Concluding remarks
  by Per Oftedal (Oslo) 197

Author Index 201

Subject Index 203

MUTAGENS IN FOOD AND GUT CONTENTS

FAECAL MUTAGENS: THEIR DISCOVERY AND POSSIBLE RELEVANCE TO THE AETIOLOGY OF LARGE-BOWEL CANCER IN MAN

S. VENITT

Chester Beatty Research Institute, Institute of Cancer Research: Royal Hospital, Pollards Wood Research Station, Nightingales Lane, Chalfont St. Giles, Buckinghamshire HP8 4SP (Great Britain)

Introduction

The epidemiology of large-bowel cancer (colorectal cancer) has been extensively reviewed by Correa and Haenszel (1978). In summary, a high incidence of large-bowel cancer is associated with consumption of a high-fat, high-protein diet, characteristic of affluent countries in North-West Europe, U.S.A. and Australia, and with urban life, whereas large-bowel cancer is relatively rare in Africa, Asia and the Andes, where average income is very low and the diet is concomitantly low in fat and protein. Studies of migrant populations have shown that racial/genetic differences in these disparate populations make little or no contribution to the observed distribution of large-bowel cancer. It is generally accepted, therefore, that diet has a large part to play in the aetiology of colorectal cancer. The content of the diet has, of course, a marked influence on the contents of the large bowel, and the possibility arises therefore that a high incidence of colorectal cancer might be associated with the presence of chemicals carcinogenic to the epithelium lining the gut. This idea is now susceptible to direct test with the advent of reliable short-term tests for potential carcinogens, most of which are based on the now well-established qualitative relationship between carcinogenicity and mutagenicity (Venitt, 1980). Several groups of investigators are now engaged in studies of the mutagenic activity of human faeces, and the present contribution is a brief summary of published data available to the author by August 1980.

Bacterial mutation studies

W.R. Bruce and his co-workers were the first to demonstrate the presence of mutagenic activity in extracts of human faeces (Bruce et al., 1977, 1979; Varghese et al., 1978). Initial studies showed that ether-extracts of freeze-dried faeces from normal healthy male volunteers, consuming a normal "Western" diet, were mutagenic to *Salmonella typhimurium* strains TA100, TA1535, TA98 without a requirement for an external source of metabolism (i.e. in the absence of S9). These and subsequent studies employed the classical plate-incorporation assay of Ames et al. (1975). Further studies using TA100 indicated that mutagenic activity was higher in ether extracts which had been washed in alkali. Fractionation of this material using high-pressure liquid chromatography revealed mutagenic activity associated with specific fractions. The level of mutagenic activity was found to differ markedly from sample to sample, and from individual to individual. Chemical investigations of volatile material from the freeze-drier trap and from freeze-dried stools showed the presence of both volatile and non-volatile nitroso compounds, including nitrosodimethylamine, nitrosodiethylamine, nitrosopyrrolidine and nitrosomorpholine (Wang et al., 1978). Nitrite and nitrate have also been shown to be present in human faeces (Tannenbaum et al., 1978). Bruce et al. (1979) claimed that the mutagen isolated from human faeces had the properties of a nitroso compound associated with a lipid. It is of interest to note that the nitrosamines detected in faeces all require metabolic activation (in the form of rodent S9) for detection of their mutagenic activity in *S. typhimurium*, whereas the mutagen partially characterised by Bruce and his co-workers was shown to be mutagenic in the absence of S9: if in fact this mutagen is a nitroso compound, it is more likely to be a nitrosamide, since nitrosamides are, in general, directly-acting alkylating agents.

Whatever the precise nature of the faecal mutagen(s) discovered by Bruce et al., the level of mutagenic activity appears to be changed by systematic changes in diet, and to differ in amount in faeces obtained from different populations. For example, Bruce et al., studied the level of faecal mutagen in one individual after the addition of either ascorbic acid or tocopherol to the diet: in both cases, there was a marked decrease in mutagenic activity following addition of either of these antioxidants to the diet, followed by a gradual increase to the control (no antioxidant) level over a period of about one month.

The second group (Ehrich et al., 1979) to have reported the presence of bacterial mutagens in human faeces employed the extraction methods of Bruce and his co-workers, and applied the technique to populations at different risk of colon cancer. Single faecal samples were collected from a group of urban whites and a group of urban blacks living in Johannesburg, and a group of blacks living in Ruskinburg, a rural locality. The age and sex-distribution of each group are summarized in Table 1. The urban whites represented a group at high risk of large-bowel cancer, the black groups representing two low-risk populations.

TABLE 1

AGE- AND SEX-DISTRIBUTION OF POPULATIONS DONATING FAECAL SAMPLES IN THE STUDY BY Ehrich et al. (1979)

Group	Male	Female	Total	Age (mean ±S.D.)
Urban whites	23	19	42	46 ± 8
Urban blacks	50	32	82	47 ± 9
Rural blacks	54	54	108	51 ± 15

Freeze-dried faeces were extracted with ether, which was then evaporated, the residue being dissolved in dimethyl sulphoxide, and tested for mutagenicity in *S. typhimurium* using the plate-incorporation method. Mutagenic activity was detected in both TA100 and TA98, and was not dependent on the addition of S9. 19% of urban-white samples were mutagenic in TA100, compared with 2% of urban black, and 0% of rural black samples, the difference between urban whites and blacks being significant at $p < 0.001$. This pattern was seen in TA98, with percentages of 10, 5 and 2 for whites, urban blacks and rural blacks respectively. This study suggests, therefore, that individuals within a group (urban whites) at high risk for large-bowel cancer were more likely to have faecal mutagens than individuals (urban blacks and rural blacks) well matched for age and sex, considered to be at lower risk of this disease. However, this conclusion is based on just one faecal sample per individual: bearing in mind the rather large variations in the level of faecal mutagens from day to day seen in the results of Bruce and his co-workers, it would at this stage seem prudent to await further confirmation of the work of Ehrich et al., before coming to far-reaching conclusions about the precise role of faecal mutagens in the aetiology of large-bowel cancer. Further work from Wilkins' laboratory on the characterization of faecal mutagens has recently been published (Lederman et al., 1980).

In summary, anaerobic incubation of faeces for 96 h prior to extraction substantially increased the mutagenicity of faecal extracts compared with non-anaerobically-incubated faeces. Incubation in the cold, in air, with antimicrobial agents, or sterilization of faeces with heat or γ-irradiation did not enhance mutagenicity. All this evidence suggests the involvement of the faecal flora in the formation of the mutagen. Thin-layer chromatography (TLC) of the TA100-positive material revealed bands fluorescing in long-wave UV, and with characteristic UV-absorption. Mutagen from several donors had the same UV-absorption and retention time on HPLC columns, and the amount of this material correlated with the level of mutagenicity. "Purified" mutagen had the same R_f on TLC as mutagen directly extracted from fresh faeces, and the faeces obtained from 5 donors all contained this material. No firm conclusions have been drawn as to the chemical nature of the purified mutagen.

A third group (Reddy et al., 1980) have recently published an investigation of faecal mutagens from groups representing populations at low or high risk of colon cancer, again using the extraction methods of Bruce and co-workers: ether extracts were partially purified by silica-gel chromatography and assayed by the standard procedures of Ames et al. (1975), using *S. typhimurium* TA98 and TA100, and Aroclor-induced rat-liver S9. The study-population consisted of 44 healthy men, of mean age 49 ± 5 years. 11 were Seventh-Day Adventists, resident in New York, consuming a vegetarian diet (milk and milk products, but no fish, fowl, or flesh) for more than 10 years and considered to represent a population at low risk of colon cancer. 15 were inhabitants of Kuopio, in rural Finland, an area whose male population suffers an age-adjusted incidence of colon cancer of 5.6 per 100 000, compared with 28.5 per 100 000 for the white male population of the U.S.A. The third group, representing this high-risk population, consisted of 18 New Yorkers. The Finns consumed more milk, dairy products and fibre, and less meat-fat than the New Yorkers: all 3 groups consumed much the same amounts of total protein, and fat-intake was lowest in the Seventh-Day Adventists. Faeces were collected over a 48-h period, pooled and extracted as described above. In this study, a sample was declared mutagenic if the ratio of the number of mutants per plate on treated plates to the number on control plates was equal to or greater than 3. Ehrich et al. (1979) considered a ratio of 2 or more indicative of mutagenicity in their study.

None of the samples from Seventh-Day Adventists showed mutagenic activity by the criterion adopted by the investigators: 13% of Finnish faecal samples were mutagenic, this activity being confined to TA98 + S9. The New Yorkers had the most mutagenic faeces, 22% of samples being active in at least one test system (i.e. one strain, with or without S9): the highest mutagenic ratios were detected in TA98 without S9, followed by TA100 without S9, then TA100 with S9. All samples active in TA100 without S9 were also active in TA98 without S9. It was concluded that faecal extracts from New Yorkers (high fat, high meat, low fibre diet) were more mutagenic than samples from the Finns (high fat, high fibre diet).

This study is the second to demonstrate marked differences in the bacterial mutagenicity of faecal extracts from groups representing populations at differing risk of large-bowel cancer. The picture which emerges is complex, the differences in strain-specificity and requirement for metabolic activation between the New Yorkers and Finns suggesting the presence of at least two and probably more classes of mutagen. The percentage of positive samples for the "high-risk" group (22%) in the study of Reddy et al. agrees closely with the 19% positive found by Ehrich et al. for their "high-risk" group, but the pattern of strain-specificity and metabolic requirement is different in the two studies, again suggesting the production of several mutagenic classes.

A study of faecal mutagens in relation to large-bowel cancer in the United Kingdom was started in September 1979, by S. Venitt, S.P. Pickering, M.J. Hill and F. Fernandez. Results obtained in this study have not been published and a brief account of our preliminary results are therefore included. Our initial studies are based on the extraction procedures pioneered by Bruce et al., and we initially examined the following extracts of freeze-dried faeces for mutagenicity, using plate-incorporation tests with several strains of *S. typhimurium* and *E. coli*, and fluctuation tests using *S. typhimurium* TA100 and *E. coli* WP2uvrA(P) (Venitt and Crofton-Sleigh, 1979; Gatehouse, 1978): (a) water from the freeze-drier trap; (b) ethyl acetate extract of (a); (c) chloroform/methanol; (d) ether extract; (e) concentrated ether extract; and (f) alkali-washed ether extract. We have examined faeces from 7 healthy individuals consuming a "Western" diet, and have observed mutagenic activity in 4 of these subjects. The extracts giving the most consistent results were (e) and (f), although mutagenic activity was occasionally seen in (c) and (d). The bacteria most sensitive to faecal mutagens were *S. typhimurium* TA100, and *E. coli* WP2uvrA(P), the most sensitive method of assay being the fluctuation test. Doses of extract giving positive results were very similar to those reported by Bruce et al., by Ehrich et al., and by Reddy et al., namely within the range 100—500 mg dry-weight of faeces per plate, in plate-incorporation assays. In concordance with the results obtained by some other groups, we find that addition of an Aroclor-induced rat-liver S9 fraction has no enhancing effect on the mutagenicity of faecal extracts. Fig. 1 shows a dose–response curve for the mutagenicity of an ether extract of freeze-dried faeces in a plate-incorporation assay using *E. coli* WP2uvrA(P) in the presence or absence of S9. Fig. 2 shows results of microtitre[R] fluctuation tests, using the same bacterial strain, and a similar faecal extract. These are pooled results from 3 separate assays of the same extract; Fig. 3 shows results where the extract was tested using 3 microtitre plates per dose in the same experiment. These results indicate the fairly reasonable consistency within and between assays of the same extract. However, we, like the other investigators, have found considerable variation in the mutagenicity of faecal samples given by the same volunteer on separate occasions. Systematic investigation of this variability is now under investigation. In order to verify that the results obtained to date reflect true mutagenic effects, rather than artefacts caused, for example, by toxicity, we have characterized the mutants obtained in fluctuation tests: samples from positive and negative wells were plated on minimal medium/agar plates supplemented with glucose, and incubated. Positive wells yielded a variety of revertant types of widely differing growth rates, which were viable upon further replating on minimal medium. The vast majority of negative wells produced to visible colonies, the occasional large colony probably representing pre-existing revertants in the original inoculum. Parallel studies using nutrient agar plates, and suitably diluted samples from negative wells allowed us to obtain information on cell survival following treatment with faecal extracts. The results so far indicate that some faecal extracts are slightly toxic at the highest doses giving mutagenic effects, and that the problem of the effect of toxicity on mutagenic yield warrants further investigation.

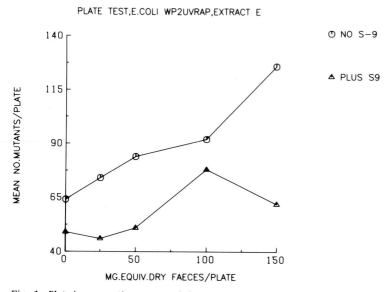

Fig. 1. Plate-incorporation assay of faecal extract (ether extract of freeze-dried stool sample), using *E. coli* WP2uvrA(P) (trp$^-$ → trp$^+$) in the presence or absence of S9 prepared from the livers of Aroclor-induced CB Hooded male rats. Each point represents the mean of 3 plates; ($p < 0.001$ for slope).

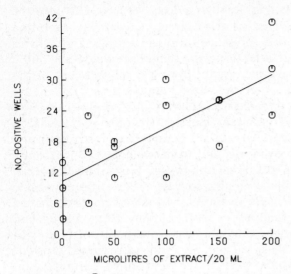

Fig. 2. Microtitre^R fluctuation test of faecal extract made by extraction of freeze-dried stool sample with diethyl ether, and the ether extract washed in 1 N Na_2CO_3. The values on the abscissa represent the volumes of extract added to 20 ml of medium prior to dispensing 200-μl samples to each of 96 wells in the microtitre^R plate. The extract was originally dissolved in dimethyl sulphoxide at a concentration equivalent to 1 g dry weight of stool per ml DMSO. The final concentrations in mg-dry-weight equivalent are therefore 0, 2.5, 5, 7.5 and 10 mg/ml respectively, or 0.5, 1, 1.5 and 2 mg per well. These are data pooled from 3 separate Expts. using the same extract. The line drawn through the points is the best fit by linear regression, $p < 0.01$.

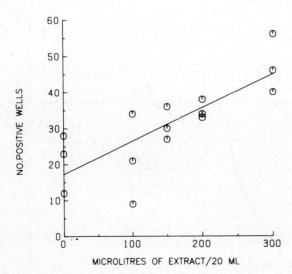

Fig. 3. Microtitre^R fluctuation test of faecal extract described in Fig. 2. These are individual data points obtained in the same experiment, using 3 plates at each dose. Doses per well are 0, 1, 2 and 3-mg-dry-weight equivalent. The line drawn through the points is the best fit by linear regression, $p < 0.001$.

Cytogenetic studies

Another approach to the study of genotoxic chemicals in human faeces has been taken by Stich and Kuhnlein (1979). These workers have used the production of chromosomal damage in cultured Chinese hamster ovary cells (CHO cells) as an indicator of mutagenic activity in human faecal extracts. Chloroform/methanol extracts of aqueous faecal slurries obtained from 3 healthy donors were evaporated to dryness, taken up in dimethyl sulphoxide (50 or 100%) and graded doses applied to CHO cells for 3 h at 37°C. The medium was then changed, and the cells incubated for a further 20 h. The frequency of metaphases containing breaks, acentric fragments and exchanges was significantly increased in cells treated with faecal extracts. The extent of chromosomal damage was dose-related, and varied between individuals and between stool samples from the same individual. Induction of chromosomal aberrations was enhanced by divalent copper or manganese ions, and inhibited by divalent or trivalent iron. Addition of catalase to the faecal extract reduced the frequency of damage. The enhancement of damage by copper and manganese and the reduction by iron and catalase suggest that the chromosome-damaging agents in the faecal extracts might be reducing agents, since these authors had previously shown that cysteine and glutathione caused chromosomal aberrations in CHO cells under similar conditions. The relevance of chromosomal damage induced aerobically in vitro by faecal extracts to the aetiology of colon cancer is discussed by Stich and Kuhnlein, who drew attention to the demonstration by Mitelman et al. (1974) and Levan et al. (1977) of abnormal chromosome complements in polyps and carcinomas of the human colon.

Conclusions

4 groups of investigators have detected mutagens in faecal extracts from a variety of subjects from widely differing communities. A 5th group has demonstrated chromosomal-damaging activity in faecal extracts. There is mounting evidence that the presence and/or levels of faecal mutagens can be modulated by changes or differences in diet, and by additions of antioxidants to the diet. These findings require confirmation and extension, and the chemical nature of the faecal mutagens, at present the subject of intensive investigation, must be clearly defined. Faeces probably contain numerous potential mutagens, bearing in mind on the one hand the enormous variety of chemicals ingested, in normal diets, in the form of food, alcoholic beverages, food additives, drugs, and contaminants, and on the other, the influence of diet on the metabolic activity and composition of the faecal flora. It is likely, therefore, that more than one chemical class will be implicated in faecal mutagenic activity.

Several candidates for the aetiological agent responsible for large-bowel cancer have been proposed, but despite considerable effort, no single agent or well defined dietary activity has been unequivocally demonstrated as a crucial determinant in the development of large-bowel cancer (see, for example, Hill, 1980; Murray et al., 1980). However, studies of the type outlined in this brief summary do allow direct investigation, in man, of at least a part of the problem. Should it be confirmed, as seems very likely, that a substantial minority of normal people (perhaps 20%), consuming what is thought of as a normal diet, nevertheless produce measurable levels of genotoxic materials in their intestinal tract, then explanations must be sought for the fact that only a fraction of these individuals will develop clinically overt malignant bowel disease. Studies of individual susceptibility (e.g. are there differences in levels of DNA-repair in the epithelial lining of the colon in different individuals?) may provide further clues.

Acknowledgement

S. Venitt acknowledges financial support from the Cancer Research Campaign and the Medical Research Council of Great Britain.

References

Ames, B.N., J. McCann and E. Yamasaki (1975) Methods for detecting carcinogens and mutagens with the Salmonella mammalian microsome mutagenicity test, Mutation Res., 31, 347—363.
Bruce, W.R., A.J. Varghese, R. Furrer and P.C. Land (1977) A mutagen in the feces of normal humans, in: Origins of human cancer, H.H. Hiatt, J.D. Watson and J.A. Winsten (Eds.), Book C, Human Risk Assessment, Cold Spring Harbor Conferences on Cell Proliferation, Vol. 4, Cold Spring Harbor Laboratory, pp. 1641—1646.
Bruce, W.R., A.J. Varghese, S. Wang and P. Dion (1979) The endogenous production of nitroso compounds in the colon and cancer at that site, in: E.C. Miller (Ed.), Naturally Occurring Carcinogens — Mutagens and Modulators of Carcinogenesis, Japan Sci. Press Tokyo/Univ. Park Press, Baltimore, pp. 221—228.
Correa, P., and W. Haenszel (1978) The epidemiology of large-bowel cancer, Adv. Cancer Res., 26, 1—141.
Ehrich, M., J.E. Aswell, R.L. van Tassell, A.R.P. Walker, N.J. Richardson and T.D. Wilkins (1979) Mutagens in the feces of 3 South-African populations at different levels of risk for colon cancer, Mutation Res., 64, 231—240.
Gatehouse, D. (1978) Detection of mutagenic derivatives of cyclophosphamide and a variety of other mutagens in a Microtitre^R fluctuation test, without microsomal activation, Mutation Res., 53, 289—296.
Hill, M.J., (1980) Bacterial metabolism and human carcinogenesis, Br. Med. Bull., 36, 89—94.
Lederman, M., R. van Tassell, S.E.H. West, M.F. Ehrich and T.D. Wilkins (1980) In vitro production of human fecal mutagen, Mutation Res., 79, 115—124.
Levan, A., G. Levan and F. Mitelman (1977) Chromosomes and cancer, Hereditas, 86, 15—30.
Mitelman, F., J. Mark, P.G. Nilson, H. Dencker, C. Norryd and K.G. Tranberg (1974). Chromosome banding pattern in human colonic polyps, Hereditas, 78, 63—68.
Murray, W.R., A. Blackwood, J.M. Trotter, K.C. Calman and C. MacKay (1980) Faecal bile acids and clostridia in the aetiology of colorectal cancer, Br. J. Cancer, 41, 923—928.
Reddy, B.S., C. Sharma, L. Darby, K. Laakso and E.L. Wynder (1980) Metabolic epidemiology of large bowel cancer, Fecal mutagens in high- and low-risk population for colon cancer, Mutation Res., 72, 511—522.
Stich, H.F., and U. Kuhnlein (1979) Chromosome breaking activity of human feces and its enhancement by transition metals, Int. J. Cancer, 24, 284—287.
Tannenbaum, S.R., D. Fett, V.R. Young, P.C. Land and W.R. Bruce (1978) Nitrite and nitrate are formed by endogenous synthesis in the human intestine, Science, 200, 1487—1489.
Varghese, A.J., P.C. Land, R. Furrer and W.R. Bruce (1978) Non-volatile N-nitroso compounds in human feces, in: E.A. Walker, L. Griciute, M. Castegnaro, R.E. Lyle and W. Davis (Eds.), Environmental aspects of N-nitroso Compounds, IARC Scientific Publications No. 19, International Agency for Research on Cancer, Lyon, pp. 257—264.
Venitt, S. (1980) Bacterial mutation as an indicator of carcinogenicity, Br. Med. Bull., 36, 57—62.
Venitt, S., and C. Crofton-Sleigh (1979) Bacterial mutagenicity tests of phenazine methosulphate and three tetrazolium salts, Mutation Res., 68, 107—116.
Wang, T., T. Kakizoe, P. Dion, R. Furrer, A.J. Varghese and W.R. Bruce (1978) Volatile nitrosamines in normal human faeces, Nature (London), 276, 280—281.

MUTAGENIC AND COLICIN-INDUCING ACTIVITY OF SOME ANTIOXIDANTS

RENANA BEN-GURION

Israel Institute for Biological Research, Ness-Ziona and Tel-Aviv University, Medical School, Tel-Aviv (Israel)

Antioxidants are commonly added to human and animal foods, to oils, to drugs and to cosmetics, mostly to prevent spoilage by oxidation. It has been reported that the addition of antioxidants to the diet can reduce the incidence of several kinds of tumours induced by certain chemicals (Wattenberg, 1972, 1973; Cumming and Walton, 1973). Wattenberg (1975) suggested that the decrease in the incidence of cancer of the stomach in the U.S.A. might be due, at least in part, to the protective effects of the antioxidants used as food additives.

So far, the actual mechanism by which the various antioxidants inhibit chemical carcinogenesis has not been elucidated. Some antioxidants can alter the properties of liver microsomes, decreasing the binding of benzo[a]pyrene (BP) metabolites to DNA, but the exact cause of the decrease has not been determined.

To study the effects of several antioxidants on the pattern of BP metabolites, mediated by microsomal enzymes, their effects were tested by assaying mutagenicity of BP metabolites in *Salmonella typhimurium* strain TA98 (Rahimtula et al., 1977). Certain antioxidants (including sodium ascorbate and pyrogallol) inhibited BP hydroxylation as well as its conversion into mutagenic metabolites by the microsomal activating mixture. It was suggested that these antioxidants exert their protective effect against cancer by inhibiting the formation of carcinogenic metabolites in the body.

Some antioxidants can reduce mutagenicity, not only by reacting with the microsomal activating mixture and thus decreasing its potential to activate premutagens into mutagens, but that they can also deactivate mutagens directly in solution. For example, sodium ascorbate can react directly with certain mutagens such as *N*-methyl-*N'*-nitro-*N*-nitrosoguanidine (MNNG) (Guttenplan, 1977) and daunomycin (Ben-Gurion, unpublished results), and cause their deactivation, resulting in the reduction of their mutagenic potential.

Antioxidants are, as a rule, reactive chemicals. It has been suggested that they have protective effects when used as food additives. But is their activity directed in one way only? Do these substances have only beneficial effects?

As far back as 1945, only one year after the discovery that DNA is the genetic material in bacteria, McCarthy (1945), one of the discoverers himself, showed that ascorbic acid can inactivate the transforming principle, the DNA in solution. Some years later, it was found that ascorbic acid can also react with viral DNA (Bode, 1957).

The interaction of ascorbic acid with viral and bacterial DNA should have raised suspicion about its possible mutagenic capability, considering the large daily intake of ascorbic acid, its addition to many food products and its potential use as an inhibitor of the intragastric formation of N-nitroso compounds and nitrosamine formation in food products. However, evidence from several experimental systems, presented only recently, has suggested that vitamin C, particularly in the presence of copper, has a mutagenic effect (Stich et al., 1976). Ascorbic acid increased the formation of metastasis when tested on epithelioma in Wistar Rats (Jacquet and Cong Hau, 1967), but such an effect is not necessarily a genetic one.

We have studied the genetic effects of some antioxidants in two test systems; the classical Ames plate test, with tester strains TA100, TA98, TA1537 and TA1535, and the colicin-induction test (Ben-Gurion, 1978), using as tester an E2 colicinogenic derivative of strain TA1537.

The colicin-induction test has some advantages over the Ames test, especially when certain toxic mutagens are tested. Some of these advantages also apply in the tests of some antioxidants. I shall describe briefly the colicin-induction test.

Colicin E2 is a specific bacteriocin, a polypeptide with antibiotic properties, produced by certain strains of bacteria and acting on other related strains. The genetic information for the synthesis of this colicin is carried on an extrachromosomal element, a plasmid. During normal growth, colicin E2 is not synthesized by the majority of the colicinogenic bacteria. Its production is probably repressed in an analogous manner to that of a temperate phage in lysogenic cells. Like the prophage in some lysogenic bacteria, colicin E2 can also be induced by treatments that cause damage to DNA or that interfere with its proper synthesis. The same mutation abolishes the capacity of the bacteria to be induced by UV in both systems (Ben-Gurion, 1976). The induction of colicin E2, like the induction of temperate phage, results in the death of the colicin-producing cell.

It is easy to measure the number of induced E2 colicinogenic cells. This was carried out by the following procedure. About 400 log-phase coliconogenic cells are mixed with about 4×10^7 indicator bacteria, which are sensitive to the specific colicin tested, in 2.5 ml of soft agar and kept at 45°C. When the soft agar containing the mixed bacteria is poured over synthetic agar plates and incubated at 37°C overnight, the indicator bacteria grow on the test plate in the form of a lawn, a uniform lawn except for small plaque-like regions. These little bald spots on the lawn of indicator cells are regions surrounding the induced colicinogenic bacteria, and their formation is the result of the inhibition of growth of the indicator bacteria. Each spot is made by one colicinogenic bacterium. These plaque-like regions are called lacunae. E2 lacunae are large, very clear and can be easily counted when assayed on synthetic plates as we have done. When the colicinogenic bacteria are irradiated before their plating, or when a substance that can damage DNA or interfere with its synthesis is added to the soft agar together with the bacteria, a certain proportion of the colicinogenic bacteria is induced on the plate and forms lacunae. The proportions of induced bacteria depends on the dose of the added mutagen, and one gets reproducible dose—response curves for various mutagens. Activating microsomal mixture can be added to the soft agar before the pouring, and thus carcinogens that have to be activated can be screened as well.

Fig. 1 demonstrates the colicin-induction test with and without an added mutagen.

Although it is too early to compare the efficiency of this test with other bacterial tests that screen for carcinogenic substances, some of the advantages

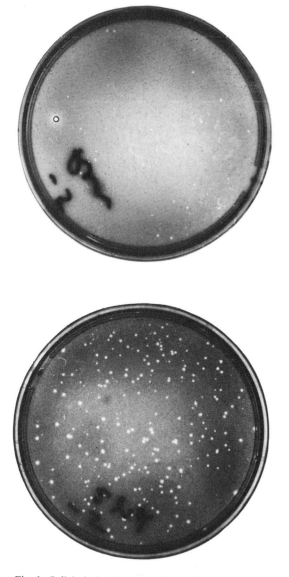

Fig. 1. Colicin-induction test plates. (1) A test plate without added mutagen. (2) A test plate with 5 μg daunomycin.

of this test can already be described. The most important advantage results from the fact that it is not necessary for the induced cell to be able to form a colony in order to produce lacuna. Some carcinogens, when tested for mutagenicity in bacteria, give false negative responses. Such negative responses depend to a large extent on the strains used as testers. Mutants can be identified, as a rule, only through their progeny — as colonies. Some toxic chemicals, which have the potential to damage DNA, are too toxic and can stop the bacteria from replicating and thus from producing progeny. These damaged cells cannot give a positive signal, because no mutant colonies appear.

Colicin production, on the other hand, is not as sensitive to many toxic effects (which are mostly active on DNA synthesis) as the capacity to grow and produce a viable colony. Many chemicals that stop DNA synthesis in the cell do not stop colicin production, which involves mainly protein synthesis.

This advantage of the colicin-induction test over the mutagenicity test can be seen in the following example, as well as in the tests of one of the antioxidants tested, to be described later. Mytomycin C, a known mutagen and also a carcinogen, is negative in the Ames test, as it kills the cells it hits. But mytomycin C is positive in the colicin-induction test. The high sensitivity of the tester strains to mytomycin C and some other mutagens results from the mutation introduced into the tester strains, the uvrB mutation. The uvrB mutation makes the cells very sensitive to many mutagenic substances, and increases the range of sensitivity to carcinogens to which they were not sensitive with the wild-type allele. This mutation also exists in the colicinogenic tester strain, and the colicinogenic tester dies after being hit by mytomycin C or some other similarly acting chemical. However, the killing of the tester does not stop the production of new lacunae induced by mytomycin C. Because the colicinogenic tester harbours the same his⁻ mutation as its parent TA1537, it can be used as a dual tester, for mutagenicity as well as for colicin-induction potential. Thus we can perform both tests with the same tester using, if needed, the parent strain TA1537 as an additional control.

There are some additional advantages of the colicin-induction test when compared with the Ames test, as follows.

(1) One tester strain is sufficient in scoring substances in the colicin-induction test for which several tester strains are needed in the Ames test, because the induction of colicin is not as specific as the reverse mutation in Ames tester strains.

(2) There are no phenocopies that can complicate the results as in the Ames test.

(3) It is also possible to test non-sterile material, since the scoring of lacunae is not very sensitive to contaminants.

Several antioxidants were tested in both tests. Some of the antioxidants were found to have colicin-inducing activity as well as a mutagenic one. The 3 antioxidants showing significant genetic activity were ascorbic acid (or sodium ascorbate), pyrogallol and purpurogallin. The genetic activity of ascorbic acid was reported in 1976 by Stich et al. (1976). These investigators found that a short exposure to a mixture of copper and ascorbic acid was mutagenic for TA100. Ascorbic acid alone did not exert a detectable mutagenic effect in their studies. They also found that cultured human fibroblasts exhibited DNA fragmentation and chromosomal abberations when treated with a mixture of ascorbic acid and copper.

We have tested ascorbic acid as well as sodium ascorbate: both substances induced colicin even without the addition of copper. However, the addition of $CuSO_4$ increased the amount of induction and reduced the minimal dose at which these substances induced colicin (Table 1). Copper alone had no detectable inducing effect. The effective dose of ascorbic acid that induced colicin was much lower than the minimal dose used by Stich et al. (1976) which, when added with copper, induced mutations.

Pyrogallol, and its oxidative derivative purpurogallin, were inducers of colicin E2 as well as being mutagenic (Tables 2 and 3). Pyrogallol had been studied before, for mutagenicity (Rahimtula et al., 1977), and found negative. But the mutagenicity test was carried out only with one tester strain, strain TA98. When we found that pyrogallol induced colicin E2 (Table 2), its mutagenicity was tested again with all 4 tester strains: TA100, TA98, TA1537 and TA1535. Table 2 shows that pyrogallol was mutagenic for TA100 and TA1535. When rat-liver microsomal preparation was added to the test plates, the mutagenic effects of pyrogallol decreased, probably as a result of an interaction with the mixture which reduced its activity towards the bacteria.

Purpurogallin, an antioxidant added to food, was also tested in both tests. Table 3 shows that it manifested only a weak mutagenicity on the test plates. A comparison of the doses that induced colicin E2 with the doses that induced mutagenicity shows that the small mutagenicity on the test plates can be observed only at higher doses than those showing significant induction of colicin. The number of mutants was not very responsive to dose. From these results, it seems that purpurogallin might have a higher genetic and mutagenic activity which is perhaps obscured to some extent by lethal events resulting from continuous exposure to the toxic chemical in the plates. Indeed, when the

TABLE 1

COLICIN INDUCTION BY ASCORBIC ACID AND SODIUM ASCORBATE

Agent(s)	Concentration in soft agar (M)	Lacunae per 400 plated cells	
		Plate 1	Plate 2
—	—	32	35
Na ascorbate	0.2×10^{-2}	40	44
Na ascorbate + $CuSO_4$	0.2×10^{-2} 10^{-6}	120	106
Na ascorbate	0.6×10^{-2}	70	61
Na ascorbate	1.4×10^{-2}	86	88
Na ascorbate	2×10^{-2}	100	93
Na ascorbate	6×10^{-2}	37	27
Ascorbic acid	0.2×10^{-2}	50	40
Ascorbic acid + $CuSO_4$	0.2×10^{-2} 10^{-6}	113	109
Ascorbic acid	0.7×10^{-2}	44	52
Ascorbic acid	1.6×10^{-2}	76	83
Ascorbic acid	2.2×10^{-2}	86	114
Ascorbic acid	7×10^{-2}	68	73

TABLE 2

MUTAGENIC AND COLICIN-INDUCING ACTIVITY OF PYROGALLOL

Dose/plate (μg)	S9 mix	Number of lacunae per 400 plated bacteria		Revertants per plate			
		Strain REN		Strain TA100		Strain TA1537	
		Plate 1	Plate 2	Plate 1	Plate 2	Plate 1	Plate 2
—	—	33	32	210	180	8	10
—	+			166	190	12	7
600	—	7	5				
400	—	27	28				
200	—	121	119	630	702	400	380
200	+			471	450	62	60
100	—	74	98				
50	—	63	71	308	323	111	204
20	—	58	76	269	291	13	29
10	—	33	40				
5	—	34	44	184	231	7	10

TABLE 3

MUTAGENIC AND COLICIN-INDUCING ACTIVITY OF PURPUROGALLIN

Dose/plate (μg)	S9 mix	Number of lacunae per 400 plated bacteria		Revertants per plate			
		Strain REN		Strain TA100		Strain TA1537	
		Plate 1	Plate 2	Plate 1	Plate 2	Plate 1	Plate 2
—	—	28	34	222	224	8	13
—	+	26	31	203	193	9	10
600	—			272	225	26	34
600	+			300	250	22	29
500	—			231	234		
500	+			363	308		
400	—			154	261	20	28
400	+			372	390	28	30
300	—			258	266	18	11
300	+			340	372	10	6
200	—			294	276	21	9
200	+			262	356	13	3
150	+	93	120				
100	—	52	33	251	289	11	13
100	+	110	108	275	339	11	18
70	—	34	41				
70	+	112	80				
50	—	58	50				
50	+	97	98				
10	—	22	26				
10	+	50	79				

bacteria were exposed to the chemical in nutrient broth for shorter times, then diluted on the test plates, the number of mutants per survivor increased with time of exposure. A reconstruction test demonstrated that the mutants in a mixture with the parent bacteria were as sensitive to the killing effects of purpurogallin as the parent bacteria themselves, thus confirming that purpurogallin is indeed mutagenic. In purpurogallin, we have another example of a toxic chemical with which it is easier to show genetic activity by using the colicin-induction test than by using the Ames test.

In terms of carcinogenicity, it is impossible, as we all know, to predict whether a mutagenic chemical will also be carcinogenic. But, in view of the high correlation between carcinogenicity and mutagenicity, and probably with other tests manifesting damage to DNA, it seems that the potential health hazard presented by every substance that shows some interaction with DNA requires careful re-evaluation.

References

Ben-Gurion, R. (1976) On the induction of a recombination-deficient mutant of *Escherichia coli*, Genet. Res., 9, 372—381.

Ben-Gurion, R. (1978) A simple plate test for screening colicine-inducing substrates as a tool for the detection of potential carcinogens, Mutation Res., 54, 289—295.

Bode, V.C. (1957) Single strand scisson induced in circular and linear λ DNA by the presence of dinitrothreitol and other reducing agents, J. Mol. Biol., 26, 125—129.

Cumming, R.B., and M.F. Walton (1973) Modification of the acute toxicity of mutagenic and carcinogenic chemicals in the mouse by prefeeding with antioxidants, Food Cosmet. Toxicol., 11, 547—553.

Guttenplan, J.B. (1977) Inhibition by l-ascorbate of bacterial mutagenisis induced by two N-nitroso compounds, Nature (London), 268, 368—370.

Jaquet, J., and H. Cong Hau (1967) Action de l'acide folique et de l'acide ascrobic sur l'evolution de l'epithelioma T8 du rat Wistar, C. R. Soc. Biol., 161, 1531—1537.

McCarthy, M. (1945) Reversible inactivation of the substance inducing transformation of pneumococcal types, J. Exp. Med., 81, 501—514.

Rahimtula, Anver D., P.K. Zachariah and P.J. O'Brien (1977) The effects of antioxidants on the metabolism and mutagenicity of benzo[a]pyrene in vitro, J. Biochem, 164, 473—475.

Stich, H.F., J. Karim and J. Koropatnick (1976) Mutagenic action of ascorbic acid, Nature (London), 260, 722—724.

Wattenberg, L.W. (1972) Inhibition of carcinogenic and toxic effects of polycyclic hydrocarbons by phenolic antioxidants and ethoxyquin, J. Natl. Cancer Inst., 40, 1425—1430.

Wattenberg, L.W. (1973) Inhibition of chemical carcinogens-induced pulmonary neoplasia, by butylated hydroxyanisole, J. Natl. Cancer Inst., 48, 1541—1544.

Wattenberg, L.W. (1975) Effects of dietary constituents on the metabolism of chemical carcinogens, Cancer Res., 35, 3326—3331.

ACETALDEHYDE NOT ETHANOL IS MUTAGENIC

G. OBE

Institut für Genetik, Freie Universität Berlin, Arnimallee 5—7, D-1000 Berlin 33 (F.R.G.)

Introduction

The mutagenic, carcinogenic and teratogenic effects of alcohol have been reviewed recently (Obe, 1980; Obe and Ristow, 1979; Obe et al., 1979a). In this communication I want to deal with some new results confirming that: (1) Not ethanol itself but its metabolite, acetaldehyde, is mutagenic. (2) There are correlations between the mutagenic and the cancerogenic activities of alcohol. (3) Ethanol is teratogenic by inhibiting the cellular RNA synthesis.

Mutagenicity

As compared to nonalcoholic controls, alcoholics have a significantly higher rate of structural chromosomal aberrations in their peripheral lymphocytes as revealed by analyzing the chromosomes from 48-h cultures (Obe, 1980; Obe et al., 1980). In 208 alcoholics we found: 1.80×10^{-3} chromatid interchanges (RB'), 2.30×10^{-3} dicentric chromosomes (DIC) and 0.87×10^{-3} ring chromosomes (Rings), resulting in a total of 4.97×10^{-3} exchange-type aberrations. In 50 nonalcoholic controls the respective values are: 0.70×10^{-3} RB', 0.10×10^{-3} DIC, 0.00 Rings, resulting in a total of 0.80×10^{-3} exchanges. The differences between the alcoholics and the controls are significant at a level of 0.05 (Obe et al., 1980). In the alcoholics there was no effect of age and sex on the aberration frequencies, but there was a significant correlation between the rate of exchange aberrations and the duration of the dependency on alcohol, indicating that the aberrations are related to the alcoholism (Obe et al., 1980).

The finding that alcoholics on treatment with antabus ($N = 51$, mean treatment time 2.5 years) i.e., in fact dry alcoholics have a lower aberration frequency than drinking alcoholics (statistically not significant), and cannot be distinguished statistically from the controls with respect to their frequencies of RB', indicates that the alcohol is the factor which is responsible for the elevation of the aberrations in the alcoholics (Obe et al., 1980).

Results concerning a possible chromosome-breaking activity of ethanol in vitro are controversial. Results from our own laboratory and others are in favor of alcohol being not active in inducing structural aberrations in mammalian including human cells (Obe, 1980). In our hands alcohol does not induce sister-chromatid exchanges (SCE) in human lymphocytes in vitro and in CHO cells (Obe, 1980). With the same ethanol concentrations used in our laboratory (0.5% v/v), Alvarez et al. (1980) found a slight elevation of SCEs in human lymphocytes after cultivation of the cells for 72 h in the presence of ethanol, even concentrations as low as 0.05% induced a significant elevation of the SCE frequencies. The control values are relatively low (3.93 SCEs per cell) and the ethanol treatments resulted in only slight elevations of the SCE frequencies (0.05%: 5.56 SCE/cell; 0.15%: 6.57 SCE/cell; 0.50%: 6.66 SCE/cell). The authors discuss their results in the sense of the lymphocytes having "... sufficient alcohol dehydrogenase activity to produce effective levels of acetaldehyde" (Alvarez et al., 1980). We treated cultures set up with the blood from different persons with 0.5% (v/v) ethanol using different treatment times (24, 48, 72 h) and found no elevation of the SCE frequencies (Königstein and Obe, unpublished).

By far the most ethanol is oxidized in the body to acetaldehyde and further to acetic acid. Only a very low amount of ingested ethanol is directly eliminated *via* the lungs or the urine. Acetaldehyde can be found in low amounts in the peripheral blood of people who drank alcohol (see respective articles in Majchrowicz and Noble, 1979).

Acetaldehyde is cross-linking isolated DNA (Ristow and Obe, 1978) and induces structural chromosomal aberrations in human lymphocytes *in vitro* from normal persons (Obe et al., 1978). In a blood culture from a patient with Fanconi's anemia, acetaldehyde was especially active in inducing structural chromosomal aberrations, and this confirms its cross-linking activity (Obe et al., 1979b). Acetaldehyde is inducing SCEs in human lymphocytes and in mammalian cells dose-dependently (Obe and Ristow, 1979). Acetaldehyde but not ethanol has been shown to induce mutations in *E. coli* WP2 uvrA trp$^-$ (Igali and Gazsó, 1980; Osztovics et al., 1980). These results indicate that not ethanol itself but rather its metabolite, acetaldehyde is mutagenic. In vivo experiments with mammals can also be interpreted in the sense of the "acetaldehyde hypothesis". Injection (i.p.) of acetaldehyde in CBA mice (1.0 or 0.5 ml of a 10^{-4}% (v/v) solution) led to a significant elevation of the rate of SCEs in the bone marrow (Obe et al., 1979b). In Chinese hamsters i.p. injection of 0.5 mg/kg acetaldehyde induced a nearly doubling of the SCE rate in the bone marrow (Korte and Obe, 1980). Feeding experiments with ethanol revealed no chromosomal aberrations or dominant lethal mutations in Wistar and in Sprague—Dawley rats and in Swiss and CD-1 mice (Obe, 1980). Intubation of high ethanol concentrations induced dominant lethals in CBA mice (Badr and Badr, 1975). Feeding of 10 or 20% (v/v) ethanol for 3—16 weeks significantly elevated the SCE rate in the bone marrow of CBA mice (Obe et al., 1979b). Feeding Chinese hamsters 10% (v/v) ethanol during a period of 46 weeks had no effect either on the SCE frequencies in the bone marrow nor on the rate of structural chromosomal aberrations in the peripheral blood lymphocytes (Korte and Obe, 1980). Tates et al. (1980) found no elevation of the frequencies of cells with micronuclei in hepatocytes and in the bone marrow, and no elevation of the SCE frequencies in the bone marrow, after feeding adult

Wistar rats 10 or 20% (v/v) ethanol as their only liquid supply for 3—6 weeks, but they found a significant elevation of the SCE frequencies in the peripheral lymphocytes. The results of Tates et al. (1980) with rats can be interpreted in the sense of acetaldehyde induced lesions being accumulated in the peripheral lymphocytes and leading to SCEs in the stimulated cells *in vitro*. In animals ingestion of ethanol will result in the formation of acetaldehyde in the body which is potentially mutagenic. It will depend on the effectiveness of the acetaldehyde metabolism whether a genetic effect can be observed or not. The Chinese hamster seems to be especially effective in metabolizing acetaldehyde, CBA mice not. The latter strain of mice seems to be well suited for analyzing the mutagenic activity of alcohol in vivo.

The mutagenic effect in man is the result of a consumption of ethanol for many years. In the sense of the acetaldehyde hypothesis it can be interpreted as resulting from the accumulation of acetaldehyde induced lesions in the long living lymphocytes of the peripheral blood (Obe et al., 1980).

Cancerogenicity

There is ample epidemiological evidence that alcohol induces cancer in man, especially of the mouth region, larynx, pharynx and esophagus (Alcohol and Cancer, 1979; Obe and Ristow, 1979; Tuyns, 1978; Rothman, 1975). Rothman (1975) on the basis of the cancer deaths in 1968 among males in the United States, calculates that about 7% of the cancers were related to alcohol consumption (see also Tuyns, 1979). The frequency of cancer of the upper digestive and upper respiratory tract is estimated to be 2.3 up to 7.2 times higher in alcoholics as compared to nonalcoholics (Schmidt and Popham, 1975/76). There is a clear synergism between alcohol consumption and tobacco smoking with respect to the risk of getting cancer of the oral cavity and larynx (McCoy and Wynder, 1979) and the esophagus (Tuyns, 1978; Smoking and Health, 1979). As Tuyns (1979) puts it "...if French males in Brittany smoked and drank as little as do their spouses, they would probably enjoy the same esophageal cancer rate, which is 20 times lower than their own".

Our chromosome analyses with alcoholics are in accord with these cancer data. Alcoholics who are smokers ($N = 152$) have a significantly higher frequency of exchange-type aberrations than alcoholics who are nonsmokers ($N = 33$) (Obe et al., 1980). Martha J. Radike (personal communication, 1980) has evidence that alcohol is cancerogenic in Sprague—Dawley rats, and she was able to show a clear synergism between alcohol ingestion and inhaled vinyl chloride with respect to cancer induction in the same strain of rats (Radike et al., 1977). With respect to the correlation between mutagenic and cancerogenic activities it may be hypothesized that ethanol is carcinogenic *via* its metabolite, acetaldehyde. Probably an acetaldehyde hypothesis can be formulated for the cancerogenic activities of alcohol also.

Teratogenicity

Alcohol is a major teratogenic agent in man, leading to the fetal alcohol syndrome (FAS) or the alcohol embryopathia (Majewski, 1979; Obe and

Ristow, 1979; Smith, 1980). The main symptoms of the FAS are intrauterine and postnatal growth retardation with facial dysmorphias, microcephalus and mental retardation. The FAS is the most frequent embryonic damage induced by an environmental agent.

We found that alcohols and aldehydes are inhibiting cellular and cell-free RNA synthesis, and there is evidence that ethanol is inhibiting the RNA synthesis in vivo (Obe and Ristow, 1979). Rawat (1976) and Obe et al. (1977) came independently to the conclusion that this activity of alcohol may be an etiological factor in the development of the FAS. Recent experiments by Brown et al. (1979) seem to be in accord with this. These authors explanted 9.5 days old conceptuses of Charles River rats within the yolk sac and amnion and cultured them for 48 h in the presence of alcohol concentrations of 150 and 300 mg per 100 ml. There was a dose-dependent reduction in embryonic growth, total DNA content and total protein content. The embryonic differentiation was clearly retarded, the alcohol-treated embryos showed a microcephalic growth. The embryos treated with 300 mg per 100 ml alcohol were retarded in growth and development for 5—7 h of gestation and had a deficiency of 8.9×10^5 cells. Cell size was not affected. The authors discuss their results in the sense that the FAS may not be produced by maternal metabolites, because it has been shown that embryos have nearly no ethanol-metabolizing activities (Obe and Ristow, 1979).

Conclusion

In conclusion it can be said that ethanol is mutagenic and cancerogenic *via* its first metabolite, acetaldehyde. Ethanol is teratogenic by inhibiting the RNA synthesis in the developing embryo and by this it causes a generalized growth retardation.

Especially with respect to the mutagenic and cancerogenic activities of alcohol in man, different congeners in the alcoholic beverages and their metabolic products as well as physiological disturbances caused by the alcohol consumption may well be of importance (Obe and Ristow, 1979).

Note added in proof

Recently it has been shown that after evaporation, alcoholic beverages contain compounds which are mutagenic in the Ames test (Loquet et al., Mutation Res., 88 (1981) 155—164; Nagao et al., Mutation Res., 88 (1981) 147—154).

References

Alcohol and Cancer Workshop (1979) Cancer Res., 39, 2815—2908.
Alvarez, M.R., L.E. Cimino Jr., M.J. Cory and R.E. Gordon (1980) Ethanol induction of sister chromatid exchanges in human cells *in vitro*, Cytogenet. Cell Genet., 27, 66—69.
Badr, F.M., and R.S. Badr (1975) Induction of dominant lethal mutation in male mice by ethyl alcohol, Nature (London), 253, 134—136.
Brown, N.A., E.H. Goulding and S. Fabro (1979) Ethanol embryotoxicity: direct effects on mammalian embryos *in vitro*, Science, 206, 573—575.
Igali, S., and L. Gazsó (1980) Mutagenic effect of alcohol and acetaldehyde on *Escherichia coli*, Mutation Res., 74, 209—210 (Abstr.).

Korte, A., and G. Obe (1980) Influence of chronic ethanol uptake and acute acetaldehyde treatment on the chromosomes of bone-marrow cells and peripheral lymphocytes of Chinese hamsters, Mutation Res., 88, 389—395.

Majchrowicz, E., and E.P. Noble (Eds.) (1979) Biochemistry and Pharmacology of Ethanol, Vols. 1 and 2, Plenum, New York.

Majewski, F. (1979) Die Alkoholembryopathie: Fakten und Hypothesen, Adv. Int. Med. Pediat. (Ergebn. Inn. Med. Kinderheilk.), 43, 1—55.

McCoy, G.D., and E.L. Wynder (1979) Etiological and preventive implications in alcohol carcinogenesis, Cancer Res., 39, 2844—2850.

Obe, G. (1980) Mutagenic activity of ethanol, in: K. Eriksson, J.D. Sinclair and K. Kiianmaa (Eds.), Animal Models in Alcohol Research, Academic Press, London, pp. 377—391.

Obe, G., and H. Ristow (1979) Mutagenic, cancerogenic and teratogenic effects of alcohol, Mutation Res., 65, 229—259.

Obe, G., H. Ristow and J. Herha (1977) Chromosomal damage by alcohol in vitro and in vivo, in: M.M. Gross (Ed.), Alcohol Intoxication and Withdrawal, Vol. IIIa, Biological Aspects of Ethanol, Plenum, New York, pp. 47—70.

Obe, G., H. Ristow and J. Herha (1978) Mutagenic activity of alcohol in man, in: Conference on Mutations, Their Origin, Nature and Potential Relevance to Genetic Risk in Man, Deutsche Forschungsgemeinschaft, H. Boldt, Boppard, pp. 151—161.

Obe, G., H. Ristow and J. Herha (1979a) Effect of ethanol on chromosomal structure and function, in: E. Majchrowicz and E.P. Noble (Eds.), Biochemistry and Pharmacology of Ethanol, Vol. I, Plenum, New York, pp. 659—676.

Obe, G., A.T. Natarajan, M. Meyers and A. den Hertog (1979b) Induction of chromosomal aberrations in peripheral lymphocytes of human blood in vitro, and of SCEs in bone-marrow cells of mice in vivo by ethanol and its metabolite acetaldehyde, Mutation Res., 68, 291—294.

Obe, G., D. Göbel, H. Engeln, J. Herha and A.T. Natarajan (1980) Chromosomal aberrations in peripheral lymphocytes of alcoholics, Mutation Res., 73 (1980) 377—386.

Osztovics, M., S. Igali, A. Antal and P. Véghelyi (1980) Alcohol is not mutagenic, Mutation Res., 74, 247 (Abstr.).

Radike, M.J., K.L. Stemmer, P.G. Brown, E. Larson and E. Bingham (1977) Effect of ethanol and vinyl chloride on the induction of liver tumors: preliminary report, Environ. Health Perspect., 21, 153—155.

Rawat, A.K. (1976) Effect of maternal ethanol consumption on foetal and neonatal rat hepatic protein synthesis, Biochem. J., 160, 653—661.

Ristow, H., and G. Obe (1978) Acetaldehyde induces cross-links in DNA and causes sister-chromatid exchanges in human cells, Mutation Res., 58, 115—119.

Rothman, K.J. (1975) Alcohol, in J.F. Fraumeni Jr. (Ed.), Persons at High Risk of Cancer, An Approach to Cancer Etiology and Control, Academic Press, New York, pp. 139—150.

Schmidt, W., and R.E. Popham (1975/1976) Heavy alcohol consumption and physical health problems: a review of the epidemiological evidence, Drug Alc. Depend., 1, 27—50.

Smith, D.W. (1980) Alcohol effects on the fetus, in: R.H. Schwarz and S.J. Yaffe (Eds.), Drug and Chemical Risks to the Fetus and Newborn, Liss, New York, pp. 73—82.

Smoking and Health (1979) A Report of the Surgeon General, U.S. Department of Health, Education and Welfare, Publ. 79-50066.

Tates, A.D., N. de Vogel and I. Neuteboom (1980) Cytogenetic effects in hepatocytes, bone-marrow cells and blood lymphocytes of rats exposed to ethanol in the drinking water, Mutation Res., 79, 285—288.

Tuyns, A. (1978) Alcohol et Cancer, Centre International de Recherche sur le Cancer, Monographie, Lyon, 42 pp.

Tuyns, A.J. (1979) Epidemiology of alcohol and cancer, Cancer Res., 39, 2840—2843.

Progress in Mutation Research, Vol 2
A. Kappas, editor.

Penicillium roqueforti TOXIN

S. MOREAU

Institut National de la Santé et de la Recherche Médicale U 42 59650 Villeneuve d'Ascq (France)

Kanota (1970) reported the toxicity of chloroform extracts from *Penicillium roqueforti* culture medium. However, he did not succeed in isolating any of the several toxic compounds. Wei et al. (1973, 1975) isolated and characterized a toxic metabolite elaborated by this Penicillium, the *Penicillium roqueforti* toxin (PRT).

In this paper I shall first examine the chemistry of *Penicillium roqueforti* metabolites, and then I shall review the biological activities of these compounds.

Structure of *Penicillium roqueforti* metabolites

PRT is a lipidic compound of molecular weight 320 which has various functional groups: an aldehyde, an α,β-unsaturated ketone, 2 epoxides and an acetyl group (Fig. 1).

In their report, Wei et al. (1975) did not mention the stereochemistry of this molecule. To study this point, we (Baert et al., 1980) carried out an X-ray diffraction analysis of a single crystal. The absolute stereochemistry, shown in Fig. 2, is typical of a sesquiterpene and more precisely the eremophilane series. The 14 and 15 methyl groups are *cis* to each other and of β configuration. The expoxides are also of β configuration.

We also looked for other metabolites in the culture medium of various strains of *P. roqueforti* used in the ripening of cheeses. Fig. 3 shows the various metabolites that we isolated: PRT and a series of metabolites that we called eremofortins A, B, C, D and E. The carbon skeletons of all the metabolites are similar.

Eremofortin A (EA) only differs from PRT at carbon 12. Eremofortin B

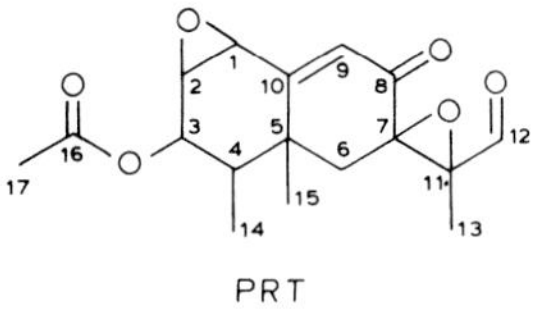

Fig. 1. Structure of PRT.

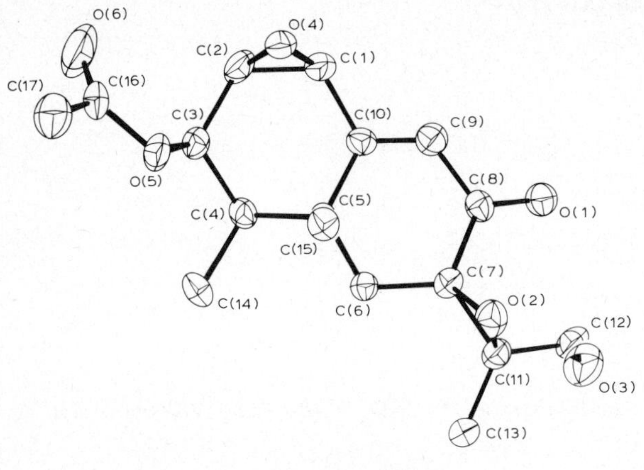

Fig. 2. The absolute stereochemistry of PRT.

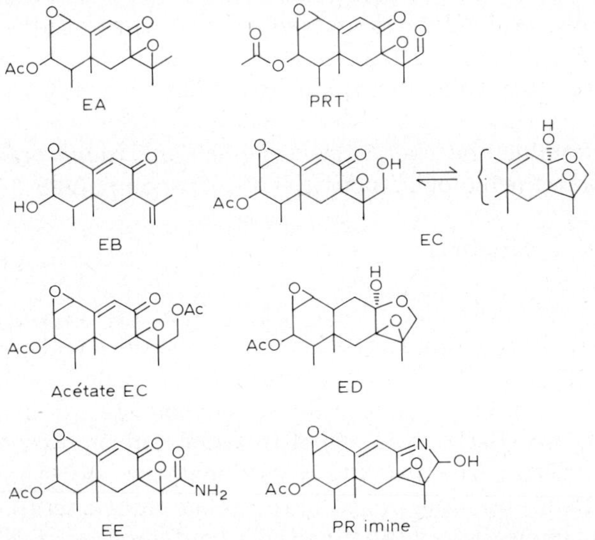

Fig. 3. Various sesquiterpenic metabolites from *P. roqueforti*.

(EB) is characterized by an isopropenyl group at position 7 and a hydroxyl group at position 4. Eremofortin C (EC) is the reduction compound of the aldehydic function at position 12 of PRT; it exists in an equilibrium between an open alcoholic form and a hemi-acetalic form. The alcoholic form represents 60% at 20°C in a solvent such as pyridine. Eremofortin D (ED) is the reduction product of the 9—10 double bond of the previous acetalic form. Eremofortin E (EE) and PR imine are metabolites that were isolated during a culture of a PRT-producing strain. During this experiment we did not obtain the usual metabolites but only these two compounds. EE is the amide derivative of PRT, while PR imine is the derivative obtained by the reaction of ammonium hydroxide on PR toxin (Moreau et al., 1976, 1977).

X-Ray diffraction studies, rotatory dispersion and circular dichroism measurements showed that all these metabolites have the same stereochemistry as PRT (Moreau et al., 1980).

Biological activities

Wei et al. (1973), in their preliminary study, mentioned an LD 50 for rats of 11 mg/kg by intraperitoneal injection and of 115 mg/kg by oral administration. However, we obtained slightly lower values of 7 mg/kg for the LD 50 by i.p. injection.

Metabolic alterations

PRT induces alterations of cell metabolism at the level of its essential mechanisms, which are related to the transmission of biological information, replication, transcription and translation.

Interaction of PRT with replication
Aujard et al. (1979) from Villejuif showed that PRT inhibits DNA biosynthesis in liver-cell cultures. This inhibition is irreversible, that is to say that the phenomenon continues after withdrawal from the incubation medium.

Interaction with transcription
Moulé et al. (1976) found that the transcription process is inhibited by PRT in a test in vitro with nuclei isolated from liver cells. By using a system, in vitro, consisting of calf-thymus DNA and bacterial RNA polymerase, they showed that the inhibition was caused by alteration of the bacterial polymerase itself and that the initiation and elongation steps of the reaction were inhibited. With this system, they also found that the activity of PRT did not require any enzymatic activation for exerting its action.

Interaction with translation
Moulé et al. (1978) showed that PRT inhibits translation in vivo. Using a system consisting of rat-liver polysomes and pH 5 enzyme fraction in vitro, these authors suggested that PRT impairs translation in vitro by altering some components of the pH 5 enzyme fraction rather than by impairment of polysomes themselves.

Structure—activity relationships
The occurrence of different PRT structurally related metabolites enables us to investigate the relationships between the chemical structure of these compounds and their respective biological properties. We therefore tested the biological activities of compounds EA, EB, EC, PR imine and 3-hydroxy derivative of PRT (Fig. 4).

For this purpose we used 3 test systems: toxicity on mice in vivo, inhibition of transcription in vitro, and inhibition of translation in vitro.

Table 1 summarizes the results (Moulé et al., 1977). PRT and its 3-hydroxy derivative are the only molecules that are biologically active. We observed a good correlation between the responses for toxicity in vivo and that obtained

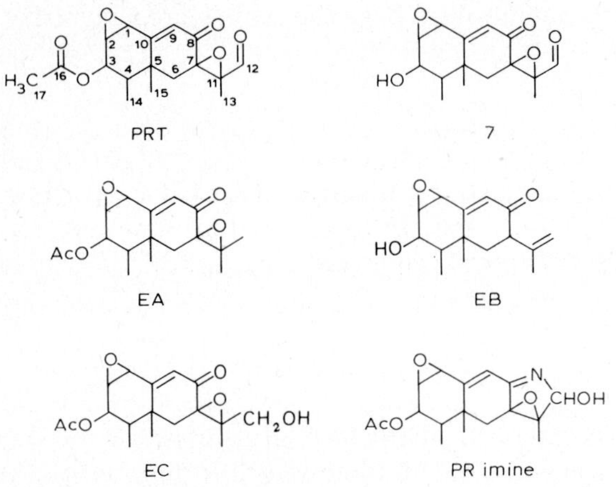

Fig. 4. Compounds tested for biological activity.

TABLE 1

BIOLOGICAL ACTIVITY OF *P. roqueforti* METABOLITES

Metabolites	Toxicity in mice		Transcription		Translation	
	mg/kg	Lethality	µg/ml	% inhi.	µg/ml	% inhi.
EA	15	0/6	800	16	1000	9
PRT	10	5/5	40	66	200	68
EB	15	0/3	800	11	500	4
EC	50	0/5	800	5	2000	11
7	10	4/4	40	62	200	63
PR imine	15	0/4	800	17	1000	10

for effects on cell-free systems for the various metabolites in vitro. The results support the view that the toxic potency and the capacity of inhibiting transcription and translation are conditioned by a common chemical structure. They also demonstrate that the aldehydic group on position 12 is directly implicated in the biological activity of the compounds tested. The loss of this functional group determines the inactivation of the molecules. In addition, the present data suggest that the 2 epoxide radicals do not play an important role in the biological activity.

Interaction of PRT with DNA

So far the long-term effects of PRT have been hardly considered; in particular there are no definitive or consistent data available on its carcinogenic potency. Wei et al. (1979) reported mutagenic effects of the toxin in tests using *Saccharomyces cerevisiae* and *Neurospora crassa*. Ueno et al. (1978) reported positive results in the Ames test; however, the difference between the number of induced revertants and that of spontaneous mutations was too low for the results of these authors to be considered as definitive.

The study of the effects of PRT on DNA may be useful in the evaluation of its genetic toxicity. The incubation of ^{14}C-labelled PRT in the presence of liver cells enabled us to show that PRT binds to nucleic acids and proteins but at different rates. The toxin binds to RNA much more than to DNA, and the lowest values of PRT binding are always found for the proteins (Moulé et al., 1980).

The centrifugation of the material obtained from incubated liver cells or incubated isolated nuclei in a caesium chloride gradient allowed us to show that PRT induces cross-links between DNA and protein from chromatin (Fig. 5). These cross-links can be cleaved by pronase treatment, and the cleavage seems to be mediated by the aldehyde group of the molecule PRT because compounds devoid of this functional group fail to cause the phenomenon. That is so for EA and PR imine.

The induction of DNA-protein cross-links mediated by an aldehyde group has been reported for formaldehyde and the other aldehydes by various authors. Feldman (1973) proposed a mechanism such as that shown in Fig. 6 to explain such cross-links induced by formaldehyde.

The implications for the cell of this type of DNA damage are still not clearly understood. Moulé (personal communication) reinvestigated the genotoxic action of PRT by using the *Salmonella typhimurium*/microsome test and the sister-chromatid exchange assay. The results, which are not yet published, indicate that PRT is inactive in these two tests.

An aldehydic function is necessary to explain all the biological properties of PRT; but is this function the only one implicated? Do the other functions, the α,β-unsaturated ketone, or epoxide contribute to this phenomenon? Is the reactivity of the aldehyde function modified by the adjacent functional groups?

In an attempt to resolve these uncertainties we synthesized molecules bearing some of the PRT functional groups such as the aldehyde and their corresponding alcohols shown in Fig. 7.

First we tested the toxicity of these various products on mice. The aldehydes were slightly more toxic than the corresponding alcohols, but we are far from the effects of PRT which was lethal at 10 mg/kg.

Then we examined the activity of these compounds on transcription and translation in vitro. We observed that they were without effect on the trans-

Fig. 5. Cross-links between DNA and proteins induced by PRT.

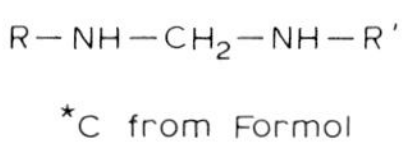

R—NH—CH$_2$—NH—R'

*C from Formol

Fig. 6. Scheme of Feldman.

Fig. 7. Synthetic aldehydes.

cription system used in vitro, whereas they evidently perturbed translation in vitro. More precisely, we observed an 80% inhibition for the aldehyde at 1000 µg/ml and a 30% inhibition for the corresponding alcohols. In this test, PRT showed a 70% inhibition at 200 µg/ml.

Finally, we examined the activity of these compounds on macromolecules, and we tested their ability to induce cross-links (Moulé et al. 1980). We observed that only the compounds bearing an aldehyde group were active but at concentrations 40 to 50 times higher than those used for PRT. A survey of the literature on the toxicity and biological activities of aldehydes showed that unsaturated aldehydes (acrolein, crotonaldehyde) are more toxic than saturated aldehydes (propionaldehydes, butyraldehydes). Formaldehyde shows an intermediate activity. Acrolein is the most toxic of all the aldehydes. It exhibits an LD50 of 6 mg/kg while crotonaldehyde is much less active (160 mg/kg). It appears, therefore, that the toxicity exhibited by PRT is exceptional. At the moment it is impossible to describe accurately the role of the other functional group in the activity of the aldehyde PRT, but this role seems obvious.

Metabolization of PRT

We have studied the metabolization of PRT by rat hepatocytes. The scheme in Fig. 8 shows the results obtained (Cacan et al., 1977). We observed 3 types of enzymatic reaction as follows. (a) A reduction of the aldehyde group on position 12 of PRT leading to EC. This reaction occurs mainly in the cytosol of the liver cell and requires NADPH. (b) A deacetylation of the acetyl group on position 3 of PRT by the microsomal fraction of the hepatocyte, which does not require NADPH. (c) The metabolization of compound EA in this system allows us to observe a hydroxylation reaction on position 12 EA. This hydroxylation reaction is classically localized on the microsomal fraction and requires molecular oxygen and NADPH.

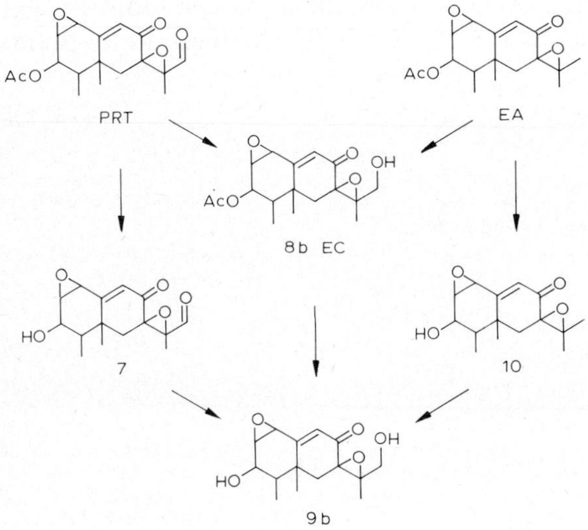

Fig. 8. Metabolization of PRT and EA by rat hepatocytes.

We can see that the reduction reaction of the aldehyde of PRT is a route of detoxification. We also found that the microsomes of rat hepatocytes contain a glucuronyl transferase able to conjugate EC (that is the reduction compound of PRT) to glucuronic acid (Cacan et al., 1978).

PR toxin and cheeses

Numerous studies have been undertaken to look for the presence of PRT in fermented cheeses. In all the studies done so far, PRT was never found in cheeses ripened by *P. roqueforti*, and indeed it cannot be found in cheese. Strains used in the ripening of blue cheeses are able to elaborate PRT on the medium used in the laboratory.

But this first and essential condition is not sufficient; the toxic strains must be able to elaborate PRT in the culture medium constituted by cheese and be able to do it during the ripening. Piva et al. (1976) showed that, in a medium with small amounts of carbohydrates, under microaerophilic conditions and in the presence of sodium chloride, *P. roqueforti* is unable to elaborate on PR toxin.

Scott and Kanhere (1979) reported that PRT was not stable in blue cheese and that the addition of PRT to a cheese led to the formation of PR imine which is not toxic as we previously showed. PR imine added to blue cheese could not be recovered over a 5-day period.

These are the essential arguments that allow us to assess the absence of toxicological hazards from PR toxin. PR toxin would never be found in cheeses.

Acknowledgements

The work reported here is the result of a collaboration between two teams. That of Dr. Y. Moulé from IRSC Villejuif for molecular biology and the team of Prof. J. Biguet from Lille (France) for cell chemical and biological works.

References

Aujard, C., E. Morel-Chany, C. Icard and G. Tringal (1979) Effects of PR toxin on liver cells in culture, Toxicology, 12, 313.

Baert, F., M. Foulon, G. Odou and S. Moreau (1980) The structure and absolute configuration of PR toxin, Acta Cryst., B36, 402.

Cacan, M., S. Moreau and R. Tailliez (1977) In vitro metabolism of *Penicillium roqueforti* toxin and a structurally related compound Eremofortin A by rat liver, Toxicology, 8, 205.

Cacan, M., S. Moreau and R. Tailliez (1978) Étude in vitro de la toxine de *Penicillium roqueforti* et de ses metabolites associés, Biochimie, 60, 685.

Feldman, M.Y. (1973) Reactions of nucleic acids and nucleoproteins with formaldehyde, Progress in Nucleic Acid Research and Molecular Biology, Vol. 13, Academic Press, New York, p. 1.

Kanota, K. (1970) Studies on toxic metabolites of *Penicillium roqueforti*, Proc. 1st US—Japan Conf. Toxic Microorganism, U.S. Dept. Intern. Unnumbered Publications, Washington DC, pp. 129—132.

Moreau, S., A. Gaudemer, A. Lablache-Combier and J. Biguet (1976) Métabolites de *Penicillium roqueforti*: PR toxines et métabolites associés, Tetrahedron Lett., 833.

Moreau, S., M. Cacan and A. Lablache-Combier (1977) Eremofortin C, a new metabolite obtained from *Penicillium roqueforti* cultures and from biotransformation of PR toxin, J. Org. Chem., 42, 2632.

Moreau, S. J. Biguet, A. Lablache-Combier, F. Baert, M. Foulon and C. Delfosse (1980) Structure et stéréochimie des sesquiterpenes de *Penicillium roqueforti* PR toxine et éremofortines A, B, C, D, E. Tetrahedron, in press.

Y. Moulé, M. Jemmali and N. Rousseau (1976) Mechanism of the inhibition of transcription by PR toxin a mycotoxin from *Penicillium roqueforti*, Chem.-Biol. Interact., 14, 207.

Moulé, Y., S. Moreau and J.F. Bousquet (1977) Relationships between the chemical structure and the biological properties of some eremophilane compounds related to PR toxin, Chem.-Biol. Interact., 17, 185.

Moulé, Y., M. Jemmali and N. Darracq (1978) Inhibition of protein synthesis by PR toxin a mycotoxin from *Penicilliom roqueforti*, FEBS Lett., 88, 341.

Moulé, Y., S. Moreau and G. Aujard (1980) Induction or cross-links between DNA and protein by PR toxin, a mycotoxin from *Penicillium roqueforti*, Mutation Res., 77, 79.

Piva, M.T., J. Guiraud, J. Crouzer and P. Galzy (1976) Influence des conditions de culture sur l'excrétion d'une mycotoxine par quelques souches de *Penicillium roqueforti*, Le Lait, 56, 397.

Scott, P.M., and S.R. Kanhere (1979) Instability of PR toxin in blue cheese, J. Assoc. Off. Anal. Chem., 62, 141.

Ueno, Y., K. Kubota, T. Ito and Y. Nakamura (1978) Mutagenicity of carcinogenic mycotoxins in *Salmonella typhimurium*, Cancer Res., 38, 536.

Wei, R.D., P.E. Still, E.B. Smalley, H.K. Schnoes and F.M. Strong (1973) Isolation and partial characterization of a mycotoxin from *Penicillium roqueforti*, Appl. Microb., 25, 111.

Wei, R.D., H.K. Schnoes, P.A. Hart and F.M. Strong (1975) The structure of PR toxin, a mycotoxin from *Penicillium roqueforti*, Tetrahedron, 31, 109.

Wei, R.D., T. Ong, W. Whong, D. Frizza, G. Bronzetti and E. Zeiger (1979) Genetic effects of PR toxin in eukaryotic microorganisms, Environ. Carcinogen., 1, 45.

Progress in Mutation Research, Vol 2
A. Kappas, editor.

BIOLOGICAL ACTIVITY OF THE AFLATOXINS

R. COLIN GARNER

Cancer Research Unit, University of York, York Y01 5DD (Great Britain)

The aflatoxins are a group of mycotoxins produced by strains of both *Aspergillus flavus* and *A. parasiticus.* They were discovered after the accidental poisoning of a large number of turkey and ducks fed contaminated peanut meal. Subsequent to this it was found that rats fed the contaminated meal developed liver cancer. Intensive investigations since these early studies have shown that there are 4 naturally occurring aflatoxins (for structures see Fig. 1) and that aflatoxin B_1 (AFB_1) is the most potent carcinogen yet discovered for the rat (Garner and Martin, 1979). Of the other 3 aflatoxins, aflatoxin G_1 (AFG_1) is about one-third as potent for liver-cancer induction, aflatoxin B_2 (AFB_2) about one-hundredth, and aflatoxin G_2 (AFG_2) non-carcinogenic. Contamination of human or animal feed by the aflatoxins occurs to a greater extent in warm humid conditions than cold temperate ones. Nevertheless, a recent report had highlighted the fact that cereal crops can be contaminated by aflatoxins even in Northern Europe (Hacking and Biggs, 1980).

The presence of aflatoxins in human foods is associated with the high human

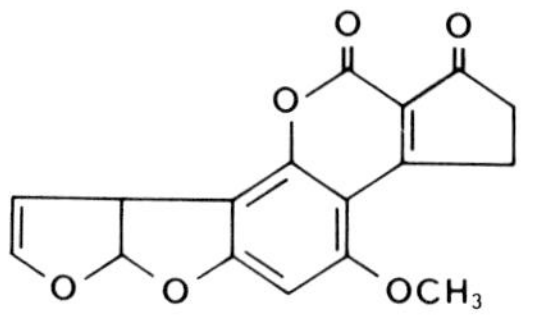

Aflatoxin B_1

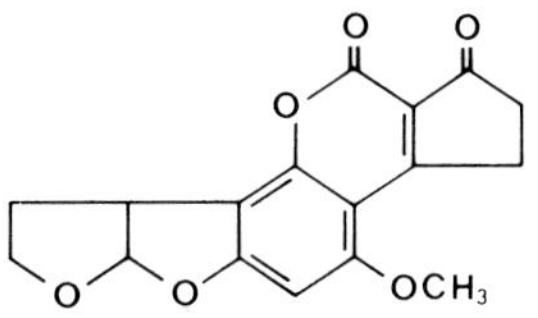

Aflatoxin B_2

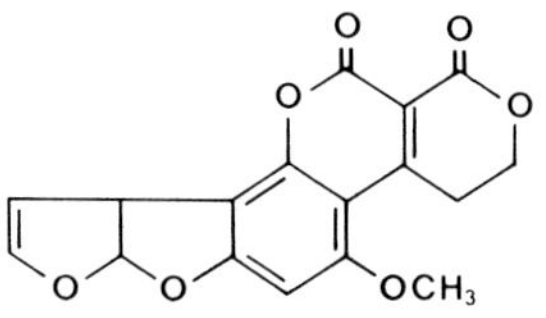

Aflatoxin G_1

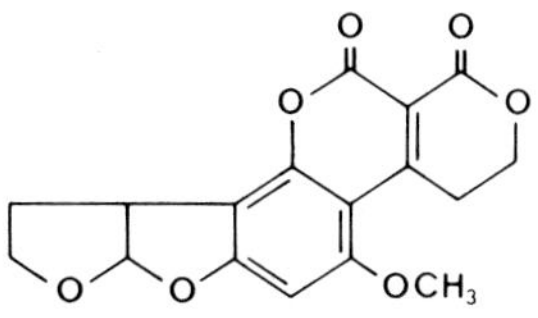

Aflatoxin G_2

Fig. 1. Structures of the naturally occurring aflatoxins.

liver-cancer incidence in certain areas of the world (Peers et al., 1976; Shank et al., 1972); recent studies suggest that there may be a synergistic action between hepatitis B virus and AFB_1 in the induction of liver cancer (Edman et al., 1980).

Animal studies have shown that AFB can induce tumours not only of the liver but also of the kidney (Ward et al., 1975), colon (Newberne and Rogers, 1973) and bone (Sieber et al., 1970). Whether tumours of other organs than the liver can be initiated in man is an open question. It should be emphasized that AFB_1 might cause tumours at sites other than the liver and that this should be borne in mind in any studies of human cancer incidence resulting from aflatoxin exposure.

The importance of AFB_1 metabolism for its carcinogenic and mutagenic effects

Work over the past 10 years in our own laboratories and several others around the world has shown that AFB_1 has to be metabolised in the body to exert its biological activity (Garner and Martin, 1979; Garner, 1980; Garner and Wright, 1975; Swenson et al., 1973, 1977). The conversion of AFB_1 to its reactive form, the 8,9-oxide, is catalysed by mono-oxygenase enzymes. The enzymes concerned are inducible by phenobarbitone but not by 3-methylcholanthrene. To date the synthesis of the 8,9-oxide has not been achieved. This electrophilic metabolite reacts readily with nucleic acids to give as the major adduct trans-8,9-dihydro-8(N^7-guanyl)-9-hydroxy AFB_1 (AFB-gua) (Martin and Garner, 1977; Essigmann et al., 1977; Lin et al., 1977). Its reaction with water yields 8,9-dihydro-9,9-dihydroxy AFB_1 (AFB_1-diol), a product formed in vitro after microsomal metabolism of AFB_1 when phosphate is used to buffer the assay medium (Neal and Cooley, 1979). In the presence of Tris-buffer a Schiff-base product is formed (Neal and Cooley, 1979; Coles et al., 1980). A summary of some of these metabolic steps is shown in Fig. 2. Other

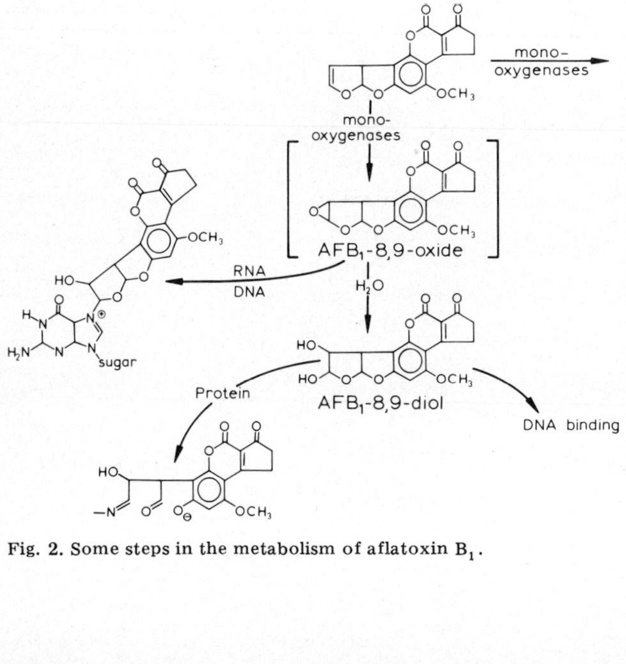

Fig. 2. Some steps in the metabolism of aflatoxin B_1.

metabolites that can be formed include aflatoxins P_1, Q, aflatoxicol, etc. Since some of these metabolites have an intact 8,9-bond it is possible that these can be further metabolised to their respective epoxides. The extent of this will be determined by, among other things, lipid solubility.

AFB_1-8,9-oxide is a major metabolite of AFB_1 in vitro. This derivative is unusual for an ultimate carcinogenic species in having poor reactivity with sulphur nucleophiles. Recent reports, however, have suggested a role for glutathione in the inactivation of the expoxide through the activity of glutathione-S-transferase (Degen and Neumann, 1978; Lotlikar et al., 1980). Studies are being carried out currently in our laboratories on the metabolism of AFB_1-diol in vivo and in vitro. This metabolite can bind to DNA and does have a weak mutagenic action (Coles et al., 1980). Whether it contributes to the overall biological activity of AFB_1 is now known.

Reactions of AFB_1-8,9-oxide with nucleic acids

AFB_1-8,9-oxide reacts readily with nucleic acids as previously mentioned. Correlations have been made between the extent of DNA reaction and biological activity. In rats, a good correlation is seen between the overall extent of DNA binding in the liver and carcinogenic susceptibility (Garner et al., 1979). Treatments which reduce the carcinogenicity of AFB_1 such as phenobarbitone also reduce levels of DNA binding (Garner, 1975). Examination of AFB_1 binding to liver DNA in the hamster, a resistant species, reveals less AFB_1 bound than in the rat. Similar correlations are found in vitro using whole cell systems such as liver slices. We have recently completed studies, in collaboration with Professor K. Norpoth, Westfalische-Wilhelms Universität, Münster (West Germany), using human-liver slices from biopsy or autopsy samples to metabolise AFB_1. Amounts of AFB_1 bound to DNA range from 4 to 27 ng/mg DNA, values between those of the mouse, a resistant species, and the rat, a susceptible species.

In a comparison of DNA binding between AFG_1 and AFB_1 in vitro and in vivo, the former carcinogen binds to about one-third the extent to DNA as AFB_1 (Garner et al., 1979). This parallels what is known about the carcinogenic potency of these two chemicals. As with AFB_1, the major DNA adduct occurs at the N^7 position of guanine. The difference in extents of DNA binding may be due to differences in the lipid solubility; AFB_1 is more lipid-soluble than AFG_1. Since other metabolites of AFB_1 such as AFM_1, AFP_1, aflatoxicol, etc. could be metabolised to reactive epoxides it is possible that some of the minor DNA adducts seen after AFB_1 administration could be due to these epoxide metabolites reacting with DNA. Furthermore, although guanine is the major base attacked by AFB_1, other bases such as adenine have been implicated in AFB_1 reactions with DNA (Garner, 1973; d'Andrea and Haseltine, 1978).

Studies on the removal of AFB_1 in vivo and in vitro

After a single administration of AFB_1 to rats (40 µg/kg) maximal binding to DNA occurs by 2 h. This is followed by a biphasic loss of radioactivity (Fig. 3) with the early phase having a half-life of some 22 h (Hertzog et al.,

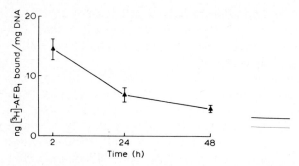

Fig. 3. Loss of AFB_1 bound to DNA in rats given a single intraperitoneal administration of 40 µg/100 g [^{3}H] AFB_1. For experimental details, see Hertzog et al. (1980).

1980). Acid hydrolysis of DNA taken from animals 2 h after AFB_1 administration shows the major DNA adduct to be AFB_1-gua. By 24 h much of this adduct has been converted to the imidazole ring-opened form 8,9-dihydro-8-(N^5-formyl-2',5',6'-triamino-4'-oxo-N^5-pyrimidyl)-9-hydroxy AFB_1 (AFB_1-triamino-Py) (Fig. 4a). Attack by OH^- at the C^8 guanine occurs because the N^7 position is ionised. We are not able to say if imidazole ring-opening occurs in vivo or during the isolation procedure. By 48 h only AFB_1-triamino-Py remains bound to the DNA (Fig. 4b).

If DNA isolated from [^{3}H]AFB_1 treated rats 2 h after carcinogen administration is incubated in vitro in a phosphate/saline buffer release of both AFB_1-diol and AFB_1gua occurs (Fig. 5). While release of the former leaves an intact guanine residue, release of the latter yields an apurinic site. Examination of the remaining AFB_1-residues bound to DNA post-incubation reveals much of the AFB_1-gua to have been converted to AFB_1-triamino-Py. By 48 h the conversion of bound adducts to this product is complete.

What are the implications of the instability of AFB_1 residues bound to DNA? AFB_1-gua is likely to be of biological consequence even though it is unstable. Its release is relatively slow in relation to cell metabolism so that bound residues may interfere with DNA replication during cell turnover. In bacteria there is evidence of a correlation between the number of AFB_1-gua residues in DNA and the induction of mutations (Stark et al., 1979). Release of AFB_1-gua from DNA will give rise to apurinic sites and these must be repaired. In bacteria error-prone repair of these could give rise to mutations. Whether similar processes occur in mammalian cells is not known. The conversion of AFB_1-gua to AFB_1-triamino-Py by OH^- attack at the C^8 position of guanine could give rise to an adduct with different properties from AFB_1-gua. The 8'-hydroxylated adduct is likely to be unstable so that imidazole ring-opening occurs. This will have the effect of converting the purine base, guanine, to a pyrimidine base as well as allowing further rotation of the aflatoxin residue around the N^7 position. This lesion appears to be more persistent than AFB_1-gua, indicating that it is not easily recognised by DNA-repair enzymes. Resistant lesions have been reported for a number of other carcinogens and it is these that might be important in tumour initiation. Finally, the rate of release of AFB_1-did appear to be pH-dependent, small differences in pH making larger differences in the rate of release (Wang and Cerutti, 1980). If there are differ-

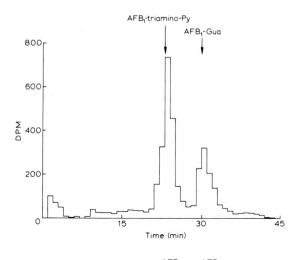

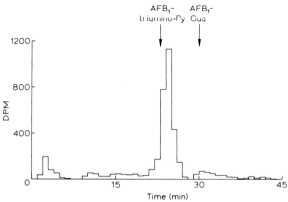

Fig. 4. Reversed-phase high-pressure liquid chromatographic profiles of liver-DNA acid hydrolysates taken from [^{3}H] AFB$_1$ treated rats: (a) 24 h and (b) 48 h after carcinogen administration.

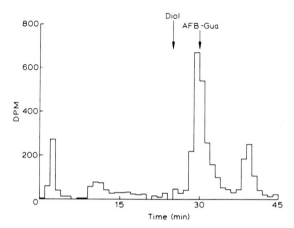

Fig. 5. Reversed-phase high-pressure liquid-chromatographic profile of the products released from rat-liver DNA isolated 2 h after [^{3}H] AFB$_1$ administration and incubated for 24 h in phosphate/saline buffer, pH 7.4.

ences in nuclear pH during for example the cell cycle, then this fact might have important consequences for tumour susceptibility.

Screening of food samples, biological fluids, etc. for the presence of aflatoxins

Early work in this area utilised silica gel thin-layer chromatography with fluorescent detection as the means of screening for aflatoxins. Comparison was made between authentic standards and the test samples in several solvent systems. These procedures had several disadvantages, in particular the lack of quantitative information which could be obtained. With the advent of high-pressure liquid chromatography it was much easier to quantitate aflatoxin contamination. Improvements in methodology such as fluorescent detection of the aflatoxins after derivitisation and the use of reversed-phase chromatography mean that as little as 1 ng of AFB_1 can be detected and measured.

Biological assays of aflatoxins have used a variety of different techniques. Up to now bacterial mutagenicity has not been used as a measure of aflatoxin contamination of foods or for the presence of aflatoxins in body fluids such as urine and faeces. We are currently embarking on such a research programme using the "fluctuation" (Forster et al., 1980) assay as a monitor of aflatoxin presence. The assay appears to be extremely sensitive, as little as 1 ng/ml of AFB_1 can be detected. Using this technique we are examining various foods commercially available in the UK such as peanuts, peanut butter, etc. A modification of the "fluctuation" assay using "microtitre" plates should improve the ease of testing.

Conclusions

The aflatoxins represent an important group of naturally occurring carcinogens to which human exposure can be limited by adequate drying and storage of food. They are an interesting group of compounds because of their biological potency, and the large species variation in tumour susceptibility. Although in both animals and man the aflatoxins initiate mainly liver cancer it should not be assumed that this is the only target organ. Evidence has been provided that human lung and colon tissue can activate AFB_1 to give DNA-bound forms (Autrup et al., 1979).

Acknowledgements

Research in our laboratories was funded by grants from the Yorkshire Cancer Research Campaign and the Medical Research Council.

References

Autrup, H., J.M. Essigmann, R.G. Croy, B.H. Trump, G.N. Wogan and C.C. Harris (1979) Metabolism of aflatoxin B_1 and identification of the major aflatoxin B_1—DNA adducts formed in cultured human bronchus and colon, Cancer Res., 39, 694—698.

Coles, B.F., A.M. Welch, P.J. Hertzog, J.R. Lindsay Smith and R.C. Garner (1980) Biological and chemical studies on 8,9-dihydroxy-8,9-dihydro aflatoxin B_1 and some of its esters, Carcinogenesis, 1, 79—90.

d'Andrea, A.D., and W.A. Haseltine (1978) Modification of DNA by aflatoxin B_1 creates alkali-labile lesions in DNA at positions of guanine and adenine, Proc. Natl. Acad. Sci. (U.S.A.), 75, 4120—4124.

Davis, N.D., and U.L. Diener (1980) Confirmatory test for the high pressure liquid chromatographic determination of aflatoxin B_1, J. Ass. Off. Anal. Chem., 63, 107—109.

Degen, G.H., and H.G. Neumann (1978) The major metabolite of aflatoxin B_1 in the rat is a glutathione conjugate, Chem.-Biol. Interact., 22, 239—255.

Edman, J.C., P. Gray, P. Valenzuela, L.B. Rall and W.J. Rutter (1980) Integration of hepatitis B virus sequences and their expression in a human hepatoma cell, Nature (London), 286, 535—538.

Essigmann, J.M., R.G. Croy, A.M. Nodzan, W.H. Busby, V.N. Reinhold, G. Buchi and G.N. Wogan (1977) Structural identification of the major DNA adduct formed by aflatoxin B_1 in vitro, Proc. Natl. Acad. Sci. (U.S.A.), 74, 1870—1874.

Forster, R., M.H.L. Green and A. Priestley (1980) Optimal levels of S9 fraction in the Ames and fluctuation tests: apparent importance of diffusion of metabolites from top agar, Carcinogenesis, 1, 337—346.

Garner, R.C. (1973) Microsome-dependent binding of aflatoxin B_1 to DNA, RNA, polyribonucleotides and protein in vitro, Chem.-Biol. Interact., 6, 125—129.

Garner, R.C. (1975) Reduction in binding of [^{14}C]aflatoxin B_1 to rat liver macromolecules by phenobarbitone pretreatment, Biochem. Pharmacol., 24, 1553—1556.

Garner, R.C. (1980) Carcinogenesis by fungal products, Br. Med. Bull., 36, 47—52.

Garner, R.C., and C.N. Martin (1979) Fungal toxins, aflatoxins and nucleic acids, in: P.L. Grover (Ed.), Chemical Carcinogens and DNA, Vol. 1, CRC Press, Fl, pp. 187—225.

Garner, R.C., and C.M. Wright (1975) Binding of [^{14}C]aflatoxin B_1 to cellular macromolecules in the rat and hamster, Chem.-Biol. Interact., 11, 123—131.

Garner, R.C., C.N. Martin, J.R. Lindsay Smith, B.F. Coles and M.R. Tolson (1979) Comparison of aflatoxin B_1 and aflatoxin G_1 binding to cellular macromolecules in vitro, in vivo and after peracid oxidation, Characterisation of the major nucleic acid adducts, Chem.-Biol. Interact., 26, 57—73.

Hacking, A., and N.R. Biggs (1980) Aflatoxin B_1 in barley in the U.K., Nature (London), 282, 128.

Hertzog, P.J., J.R. Lindsay Smith and R.C. Garner (1980) A high pressure liquid chromatography study on the removal of DNA-bound aflatoxin B_1 in rat liver and in vitro, Carcinogenesis, 1, 787—793.

Lin, J.K., J.A. Miller and E.C. Miller (1977) 2,3-Dihydro-2-(guan-7-yl)-3-hydroxyaflatoxin B_1, a major acid hydrolysis production of aflatoxin B_1—DNA or —ribosomal RNA adducts formed in hepatic microsome-mediated reactions and in rat liver in vivo, Cancer Res., 37, 4430—4438.

Lotlikar, P.D., B.M. Insetta, P.R. Lyons and E.C. Jhee (1980) Inhibition of microsome-mediated binding of aflatoxin B_1 to DNA by glutathione-S-transferase, Cancer Lett., 9, 143—149.

Martin, C.N., and R.C. Garner (1977) Aflatoxin B_1-oxide generated by chemical or enzymic oxidation of aflatoxin B_1 causes guanine substitution in nucleic acids, Nature (London), 267, 863—865.

Neal, G.E., and P.J. Cooley (1978) Some high performance liquid chromatographic studies of the metabolism of aflatoxins by rat liver microsomal preparations, Biochem. J., 174, 839—851.

Neal, G.E., and P.J. Cooley (1979) The formation of 2,3-dihydro-2,3-dihydroxy aflatoxin B_1 by the metabolism of aflatoxin B_1 in vitro by rat liver microsomes, FEBS Lett., 101, 382—386.

Newberne, P.M., and A.E. Rogers (1973) Rat colon carcinomas associated with aflatoxin and marginal vitamin A, J. Natl. Cancer Inst., 50, 439—448.

Peers, H.G., G.A. Gilman and C.A. Russell (1976) Dietary aflatoxins and human liver cancer, A study in Swaziland, Int. J. Cancer, 17, 167—176.

Shank, R.C., N. Bhamarapravati, J. Gordon and G.N. Wogan (1972) Dietary aflatoxins and human liver cancer in two municipal populations of Thailand, Food Cosmet. Toxicol., 10, 171—179.

Sieber, S.M., P. Correa, D.W. Dalgard and R.H. Adamson (1970) Induction of osteogenic sarcomas and tumours of the hepatobiliary system in nonhuman primates with aflatoxin B_1, Cancer Res., 39, 4545—4554.

Stark, A.A., J.M. Essigmann, A.L. Demain, T.R. Skopek and G.N. Wogan (1979) Aflatoxin B_1 mutagenesis, DNA binding and adduct formation in *Salmonella typhimurium*, Proc. Natl. Acad. Sci. (U.S.A.), 76, 1343—1347.

Swenson, D.H., J.A. Miller and E.C. Miller (1973) 2,3-Dihydro-2,3-dihydroxy-aflatoxin B_1: an acid hydrolysis product of an RNA—aflatoxin B_1 adduct formed by hamster and rat liver microsomes in vitro, Biochem. Biophys. Res. Commun., 83, 1260—1267.

Swenson, D.H., J.K. Lin, E.C. Miller and J.A. Miller (1977) Aflatoxin B_1-2,3-oxide as a probable intermediate in the covalent binding of aflatoxins B_1 and B_2 to rat liver DNA and ribosomal RNA in vivo, Cancer Res., 37, 172—181.

Wang, T.V., and P. Cerutti (1980) Spontaneous reactions of aflatoxin B_1 modified deoxyribonucleic acid in vitro, Biochemistry, 19, 1692—1698.

Ward, J.M., J.M. Sontag, E.K. Weisburger and C.A. Brown (1975) Effect of lifetime exposure to aflatoxin B_1 in rats, J. Natl. Cancer Inst., 55, 107—113.

THE ROLE OF BACTERIAL METABOLISM IN THE GUT IN RELATION TO LARGE BOWEL CANCER

M.H. THOMPSON and M.J. HILL

Bacterial Metabolism Research Laboratory, Central Public Health Laboratory, Colindale Avenue, London NW9 and Bacterial Metabolism Research Laboratory, P.H.L.S. Centre for Applied Microbiology and Research, Porton Down, Salisbury, Wiltshire (Great Britain)

Cancer of the large bowel is one of the major causes of cancer mortality but as yet the agents responsible for its initiation have to be identified. Epidemiological studies have demonstrated that this cancer is most prevalent in the urbanised societies of Western Europe and North America (Doll, 1969) when compared to the rural societies of the developing world. Further studies have demonstrated that environmental factors are the major contributing elements in the incidence of this disease, with the closest correlation being with variations in dietary habits (Graham and Mettlin, 1979). More detailed dietary analysis has revealed that in the high risk regions there is an enhanced consumption of fat (Wynder, 1975) and animal protein (Haenszel et al., 1973) and that the diet contains more highly processed carbohydrate (Burkitt and Trowell, 1975) and is depleted in fibre (Burkitt et al., 1972). However, these investigations have not led to the identification of carcinogenic agents in the diet responsible for the initiation of large bowel cancer, indeed direct-acting carcinogens must be eliminated as it is unlikely that they would survive the digestive tract to express their activity only in the colon. Most known carcinogens require metabolic activation to express their biological activity (Miller, 1970) but analysis of the colonic contents from a series of subjects has yet to identify positively any carcinogen or pre-carcinogen. Of the faecal components analysed it has been demonstrated that the stools of high risk populations contain higher concentrations of more highly degraded bile acids and neutral steroids when compared to those from low risk groups (Hill et al., 1971; Reddy et al., 1978). Furthermore, these stools contain higher concentrations of unsaturated long-chain fatty acids (Sperry et al., 1976). A preliminary study has also revealed that faeces may contain volatile carcinogenic N-nitroso amino derivatives (Land and Bruce, 1978).

With respect to the role of faecal steroids in the aetiology of this cancer an initial study revealed a correlation between the concentration of faecal bile acids, especially the secondary bile acid deoxycholic acid, and colon cancer incidence rates (Hill, 1975). It has also been demonstrated that in vitro faecal bacteria can metabolise bile acids via a variety of pathways, including:

(1) Bile salt deconjugation

The taurine and glycine conjugates of bile acids are readily metabolised by a wide range of faecal bacteria (Aries and Hill, 1970a). This enzymic activity is widely distributed and is found in faecal samples from all of the regions examined.

The wide distribution of this activity, which ensures that most of the faecal bile acids are present as the free acid, does not correlate with the incidence of large bowel cancer (Fig. 1).

(2) Hydroxyl dehydrogenase

Hydroxyl groups at positions 3, 7 and 12 of the cholanic acid ring may be dehydrogenated to a keto group by a wide range of faecal bacteria including enterobacteria, enterococci, clostridia, bacteroides and bifido-bacteria (Aries and Hill, 1970b). This activity, especially the formation of 7 keto derivatives, was observed in a range of subject groups but did not correlate with the cancer rates (Fig. 1).

(3) Dehydroxylation

Formation of the secondary bile acids deoxycholic acid and lithocholic acid, from cholic acid and chenodeoxycholic acid respectively, proceeds in faeces via a dehydroxylating enzyme activity. In vitro studies have partially characterised only the 7-dehydroxylase activity which requires strictly anaerobic conditions (Aries and Hill, 1970b). This activity is confined to the anaerobic bacteria that are much more prevalent in the faeces of subjects living in the higher risk environments of Western Europe and North America and therefore correlates to some degree with the cancer incidence data. Whilst this activity competes with the 7α-hydroxyl-dehydrogenase activity, under the conditions prevailing in the colon it is the dominant activity and is reflected in the excellent correlation observed between faecal deoxycholic acid concentration and large bowel cancer rates (Hill, 1975) (Fig. 1).

(4) Nuclear dehydrogenation

Δ^1 and Δ^4 nuclear dehydrogenation activities in association with 3-keto substituents have been demonstrated in a group of clostridia bacteria present in faeces (Aries et al., 1971). This inducible activity requires strict anaerobic conditions and a 5β conformation between rings A and B of the steroid nucleus. Thus introduction of a Δ^1 double bond may preceed the introduction of a Δ^4 but not vice versa. It is of particular interest that the clostridia capable of carrying out this reaction (the nuclear dehydrogenating (NDH) clostridia) are carried by up to 40% of the population from high risk areas but appear to be rare in the faeces from low risk populations (Hill, 1974) (Fig. 2).

(5) Aromatisation

In a restricted number of cases the clostridia able to introduce double bonds into the steroid nucleus may also aromatise ring A (Goddard and Hill, 1972) and, with suitable substrates in one case, ring B of the steroid nucleus (Goddard and Hill, 1973) (Fig. 2).

By analogy to the structures of known polycyclic aromatic carcinogenic

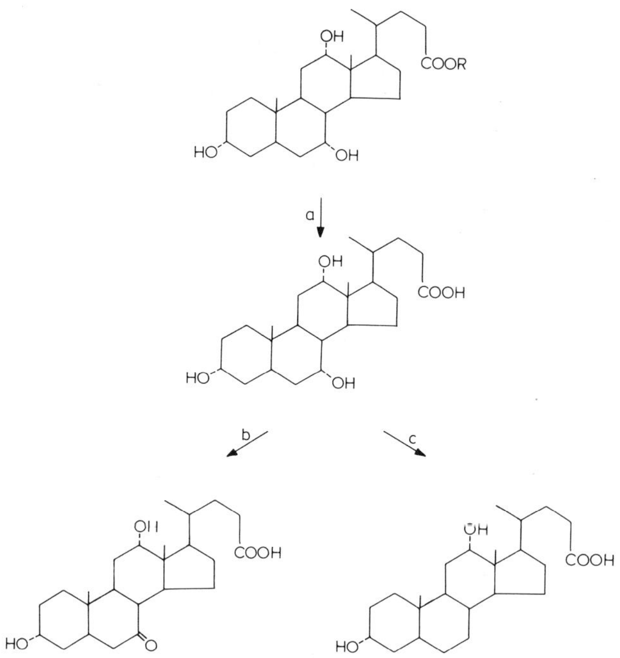

Fig. 1. Bile acid (i.e. cholic acid) metabolism mediated by faecal bacteria: (a) Deconjugation (R = taurine or glycine). (b) Hydroxy dehydrogenation. (c) Dehydroxylation.

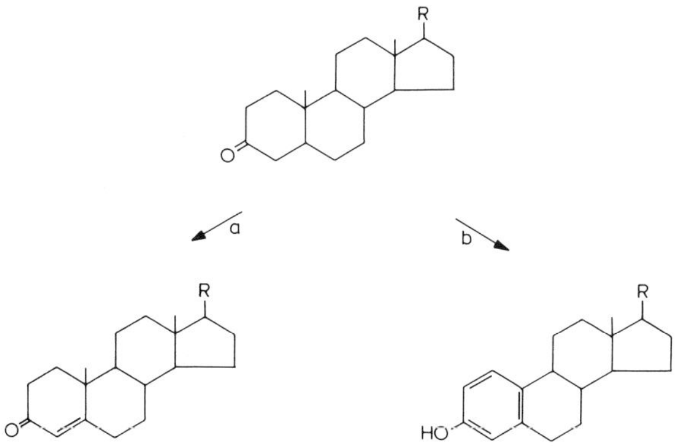

Fig. 2. Metabolism of the steroid nucleus mediated by faecal clostridia: (a) 3-oxo-Δ^4-steroid dehydrogenation (R = keto group or —CH(CH$_3$)CH$_2$ · CH$_2$ · COOH). (b) Aromatisation (R = keto group).

hydrocarbons it was originally postulated that faecal bacteria may modify bile acids by a combination of the elucidated enzyme reactions, plus some theoretical steps, to yield cyclopentaphenanthrene-like derivatives (Hill, 1975). More recent determinations of the structures of the mutagenic metabolites responsible for the activity of polycyclic hydrocarbons, such as benzo[a]pyrene (Malaveille et al., 1977) and 11-methylcyclopentaphenanthren-17-one (Coombes et al., 1979) has revealed that the biologically active centre in many cases resides in a 'diol epoxide' metabolite with a particular stereochemical conformation — the so-called 'bay region epoxide' (Wood et al., 1979). On this basis the analogy may be simplified to suggest that bile acids may be metabolised by faecal bacteria to yield substrates with structures akin to the precursors of diol epoxides. Furthermore, this metabolism may be carried out by enzymes characteristic of the faecal bacteria from high risk areas. Thus deconjugation of the tauro- and glyco-deoxycholic acid conjugates may be followed by 3α-hydroxy dehydrogenation, Δ^4 nuclear dehydrogenation and 7α-hydroxy dehydroxylation to yield 3-oxo-4,6-choladienoic acid with a potential 'bay region' structure between rings B, C and D (Fig. 3). However, the presence of such substrates in faeces remains to be demonstrated and their biological activity remains unknown.

The distribution of NDH clostridia has been assayed in a large number of specimens from different populations and patient groups, along with the faecal concentrations of bile acids. The NDH activity is assayed by selectively growing clostridia on a plate, selecting 10 colonies at random and culturing them in broth with inducer for 3 days and then with 5β-androst-4-ene-3,17-dione and menadione for at least 12 h. NDH activity is then determined by assaying by t.l.c. for the conversion of substrate to 5β-androst-4-ene-3,17-dione. The bile acid concentration is determined by the 3-hydroxy steroid dehydrogenase technique (Hill and Aries, 1971). As already indicated international studies have revealed that faecal samples from low risk regions have low faecal bile acid concentrations and carriage rates of NDH clostridia whereas samples from high risk areas have relatively high concentrations of bile acids and elevated carriage rates of NDH clostridia (Goddard et al., 1975). This joint discriminant correlates well with the large bowel cancer incidence data.

More detailed analysis of faecal samples from case-controlled groups has yielded similar data. In one study patients with conditions predisposing to colorectal cancer have been followed. Long-term patients with ulcerative colitis may develop cancer of the colon but as a group neither their faecal bile acid

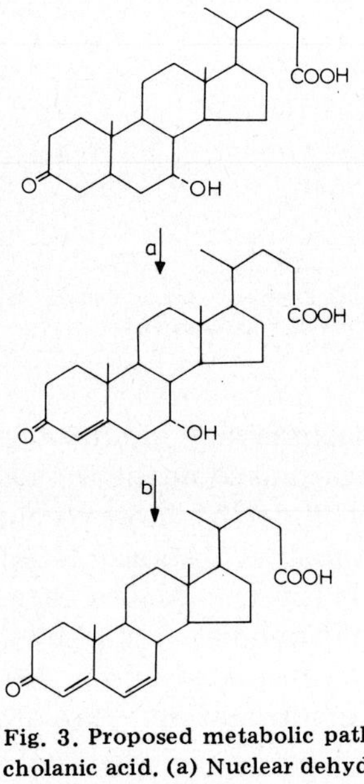

Fig. 3. Proposed metabolic pathway for the formation of 3-oxo-4,6-choladienoic acid from 3-oxo-7αOH-cholanic acid. (a) Nuclear dehydrogenation, (b) Dehydroxylation.

concentrations nor their carriage rate of NDH clostridia varied significantly from controls (Table 1). However, the limited number of patients that have developed carcinoma so far (6/79) or have exhibited moderate or severe dysplasia (7/79) have higher than average mean faecal bile acid concentrations. Furthermore, most colon carcinomas develop in adenomatous polyps, the stage at which carcinomas appear relating to the size of the adenoma. In a study of a group of patients known to carry nonmalignant polyps it was demonstrated that whilst the average faecal bile acid concentration was high there was a graduation of concentration, with the most at risk patients with the biggest polyps having the highest concentration (Table 1). The patients with the largest adenomas also had the highest NDH clostridia carriage rates. Thus in 2 studies of patients with conditions predisposing to carcinoma of the bowel a trend to higher faecal bile acid concentration and NDH clostridia carriage was observed.

In a concurrent study of patients suffering from colorectal cancer it has been possible to sub-classify the faecal bile acid concentrations and NDH clostridia carriage rates according to the site at which the tumour has developed and to the degree of invasiveness (Table 2). When compared to controls all colorectal cancer patients in this group (excluding those with the most advanced invasive tumours) have both elevated bile acid concentrations and NDH carriage rates. Of the subjects examined this carriage rate was consistent for all colon cancers regardless of the site of tumour development but those with cancer of the descending colon had significantly higher bile acid concentrations. Similarly fractionation of the carriage rate and bile acid concentration with respect to the position of the tumour in rectal cancers revealed raised carriage rates of NDH clostridia and uniformly elevated bile acid concentrations. From these studies it is not possible, with the limited numbers presently available, to correlate the NDH clostridia carriage rate/faecal bile acid concentration discriminant with the site at which colorectal tumours develop, although there is evidence that caecal, sigmoid and rectal cancers may have slightly different aetiologies (Wynder, 1975). Furthermore if these

TABLE 1

FAECAL BILE ACID CONCENTRATIONS AND CARRIAGE RATES OF NDH CLOSTRIDIA IN SUBJECTS WITH CONDITIONS PREDISPOSING TO COLORECTAL CARCINOMA

Patient group	Number of subjects	Mean faecal bile acid concentration (mg/g dry faeces)	% carrying NDH clostridia
Long-term ulcerative colitis	79	7.7	29
(i) with subsequent carcinoma	6	11.0	40
(ii) with moderate to severe dysplasia	7	10.9	38
Patients with colon polyps	72	8.1	42
Non-adenomatous polyps	15	6.6	33
Adenomas <0.5 mm diameter	12	6.0	8
6—10 mm	10	7.8	20
11—15 mm	8	8.5	50
16—20 mm	10	8.4	50
<20 mm	17	10.6	76
Normal	—	7.9	~35

TABLE 2

FAECAL BILE ACID CONCENTRATION AND CARRIAGE RATES OF NDH CLOSTRIDIA IN SAMPLES FROM SUBJECTS WITH DIAGNOSED COLORECTAL CARCINOMA

Tumour site of classification	Number of subjects	Mean faecal bile acid concentration (mg/g faeces)	% carrying NDH clostridia
Colon cancer	30	10.6	77
Caecum, ascending, transverse colon	10	9.0	77
Descending or sigmoid colon	20	11.4	80
Rectal cancer	38	10.3	76
Upper third	13	10.3	86
Mid third	12	10.8	46
Lower third	13	9.9	82
All colorectal cancers (excluding highly invasive tumours)	68	10.4	77
'Dukes' classification			
A	17	10.6	86
B	37	10.7	72
Cl^-/Cl^+	25	9.6	83
C2	6	7.7	50
Normal	—	7.9	35

colorectal cancers are examined by degree of invasiveness, whilst the NDH clostridia carriage rates are high there is an approximate inverse correlation with bile acid concentration. The latter is to some degree a reflection of metastasis with subsequent effect on liver function and bile acid turnover which may lead to reduced faecal bile acid concentration. Such subjects should be treated separately when determining faecal parameters as their faecal composition may reflect conditions other than colorectal cancer, such as impaired liver function.

In conclusion it has been demonstrated that in vitro some faecal bacteria of the clostridia species can modify bile acids to unsaturated derivatives with characteristics similar to known pre-carcinogens. Furthermore, faecal samples from high risk populations, individuals with conditions predisposing to colorectal carcinoma development and carcinoma patients have relatively high concentrations of faecal bile acids and carriage rates of the nuclear dehydrogenating clostridia. These faecal characteristics may in some measure reflect adaptation by these groups to the consumption of a diet containing elevated quantities of fat and animal protein. There are therefore indications that one of the aetiological factors associated with the incidence of large bowel cancer may be the production of carcinogens or pre-carcinogens in situ by faecal bacteria acting upon dietary secretions arising in response to dietary components.

Acknowledgments

We would like to acknowledge that this research was supported financially by the Cancer Research Campaign.

References

Aries, V., and M.J. Hill (1970a) Degradation of steroids by intestinal bacteria, I. Deconjugation of bile salts, Biochim. Biophys. Acta, 202 526—534.

Aries, V., and M.J. Hill (1970b) Degradation of steroids by intestinal bacteria, II. Enzymes catalysing the oxidoreduction of the 3α-, 7α- and 12α-hydroxyl groups in cholic acid and the dehydroxylation of the 7-hydroxyl group, Biochim. Biophys. Acta, 202, 535—543.

Aries, V., P. Goddard and M.J. Hill (1971) Degradation of steroids by intestinal bacteria, III. 3-Oxo-5β-steroid-Δ^1-dehydrogenase and 3-oxo-5β-steroid-Δ^4-dehydrogenase, Biochim. Biophys. Acta, 248, 482—488.

Burkitt, D.P., and H.C. Trowell (1975) in: Refined Carbohydrate Foods and Disease, Academic Press, New York.

Burkitt, D.P., A.R.P. Walker and N.S. Painter (1972) Effect of dietary fibre on stools and transit-times, and its role in the causation of disease. Lancet II, 1408—1411.

Coombes, M.M., A.M. Kissonerghis, J.A. Allen and C.W. Vose (1979) Identification of the proximate and ultimate forms of the carcinogen 15,16-dihydro-11-methyl-cyclopenta[a]phenanthrene-17-one, Cancer Res., 39, 4160—4165.

Doll, R. (1969) The geographical distribution of cancer, Br. J. Cancer, 23, 1—8.

Goddard, P., and M.J. Hill (1972) Degradation of steroids by intestinal bacteria, IV. The aromatisation of ring A, Biochim. Biophys. Acta, 280, 336—341.

Goddard, P., and M.J. Hill (1973) The dehydrogenation of the steroid nucleus by human-gut bacteria, Trans. Biochem. Soc., 1, 1113 1115.

Goddard, P., F. Fernandez, B. West, M.J. Hill and P. Barnes (1975) The nuclear dehydrogenation of steroids by intestinal bacteria, J. Med. Microbiol., 8, 429—435.

Graham, S., and C. Mettlin (1979) Diet and colon cancer. Am. J. Epidemiol. 109, 1—20.

Haenszel, W., J.W. Berg, M. Segi, M. Kurihara and F.B. Locke (1973) Large bowel cancer in Hawaiian Japanese, J. Natl. Cancer Inst., 51, 1765—1779.

Hill, M.J. (1974) Bacteria and the etiology of colonic cancer, Cancer, 34, 815—818.

Hill, M.J. (1975) Metabolic epidemiology of dietary factors in large bowel cancer, Cancer, 36, 2387—2400.

Hill, M.J., and V.C. Aries (1971) Faecal steroid composition and its relationship to cancer of the large bowel, J. Pathol., 104, 129—139.

Hill, M.J., B.S. Drasar, V. Aries, J.S. Crowther, G. Hawksworth and R.E.O. Williams (1971) Bacteria and aetiology of cancer of the large bowel, Lancet, 1, 95—100.

Land, P.C., and W.R. Bruce (1978) Fecal mutagens: A positive relationship with colorectal cancer, Proc. Am. Assoc. Cancer Res., 19, 167.

Malaveille, C., J. Kuroki, P. Sims, P.L. Grover and H. Bartsch (1977) Mutagenicity of isomeric diol-epoxides of benzo[a]pyrene and benz[a]anthracine in S. typhimurium TA98 and TA100 and in V79 Chinese hamster cells, Mutation Res., 44, 313—326.

Miller, J.A. (1970) Carcinogenesis by chemicals: An overview, Cancer Res., 30, 559—576.

Reddy, B.S., A.R. Hedges, A.R. Laakso and E.L. Wynder (1978) Metabolic epidemiology of large bowel cancer: Faecal bulk and constituents of high risk North American and low risk Finnish populations, Cancer, 42, 2832—2838.

Sperry, J.F., A.A. Salyers and T.D. Wilkins (1976) Long chain fatty acids and colon cancer risk, Lipids, 11, 637—639.

Wood, A.W., W. Levin, R.L. Chang, H. Yagi, D.R. Thakker, R.E. Lehr, D.M. Jerina and A.H. Conney (1979) Bay-region activation of polycyclic hydrocarbons, in: P.W. Jones and P. Leber (Eds.), Polynuclear Aromatic Hydrocarbons, Third Int. Symposium on Chemistry and Biology — Carcinogenesis and Mutagenesis. Ann Arbor Publishers, 1979, pp. 531—551.

Wynder, E.L. (1975) The epidemiology of large bowel cancer, Cancer Res., 35, 3388—3394.

MUTAGEN—CARCINOGENS IN AMINO ACID AND PROTEIN PYROLYSATES AND IN COOKED FOOD

T. MATSUSHIMA and T. SUGIMURA

Department of Molecular Oncology, Institute of Medical Science, University of Tokyo and National Cancer Center Research Institute, Tokyo (Japan)

Summary

Charred parts of broiled fish and beef showed mutagenic activity toward *Salmonella typhimurium* TA98 with metabolic activation. Recent findings on new mutagens isolated from pyrolysates of amino acids and protein foods are reviewed. We previously isolated potent mutagens, 3-amino-1,4-dimethyl-5*H*-pyrido[4,3-*b*]indole (Trp-P-1) and 3-amino-1-methyl-5*H*-pyrido[4,3-*b*]indole (Trp-P-2) from tryptophan pyrolysate and 2-amino-6-methyl-dipyrido[1,2-*a*:3',2'-*d*]imidazole (Glu-P-1) and 2-aminodipyrido[1,2-*a*:3',2'-*d*]imidazole (Glu-P-2) from glutamic acid pyrolysate. Trp-P-1 and Trp-P-2 were hepatocarcinogenic in mice. Females were more susceptible than males. Trp-P-1, Trp-P-2 and Glu-P-2 were actually found in broiled food.

Recently 2 new mutagens, 2-amino-3-methylimidazo[4,5-*f*]quinoline (IQ) and 2-amino-3,4-dimethylimidazo[4,5-*f*]quinoline (MeIQ) were isolated from broiled sardines. IQ was also found in fried beef. The specific mutagenic activities of IQ and MeIQ were higher than that of aflatoxim B_1. These new series of heterocyclic amines were activated by cytochrome P-448 from the liver of rats treated with polychlorinated biphenyls or 3-methylcholanthrene. Both in vivo and in vitro carcinogenicity experiments with these compounds, including IQ and MeIQ, are under way.

Introduction

Epidemiological studies of the incidence of cancer among migrants have clearly demonstrated that environmental factors, especially factors present in food, play an important role in causing human cancer. First generation Japanese immigrants to Hawaii showed the same high incidence of stomach cancer and low incidence of colon cancer as native Japanese. However, second and third generation immigrants showed a decreased incidence of stomach cancer with a concomitant increase in the incidence of colon cancer, equivalent to the respective levels of incidence in white people (Haenszel and Kurihara, 1968). This trend in the incidence of cancer is mainly dependent on changes in diet.

A mutagenicity test using *Salmonella typhimurium* (Ames et al., 1975; McCann et al., 1975a) has been used in screening for possible carcinogens in our daily food because there is a close correlation between mutagenicity and carcinogenicity (McCann et al., 1975b; Sugimura et al., 1976). A mutation test using *S. typhimurium* TA98 and TA100 was carried out with the preincubation technique (Matsushima et al., 1980). High mutagenic potential was detected in smoke from broiled fish (Sugimura et al., 1977b) and also in the charred surfaces of sardine and beefsteak (Nagao et al., 1977b). As shown in Table 1, smoke condensate obtained by pyrolysis of protein contained high mutagenic activity to TA98 with metabolic activation (Nagao et al., 1977a). Smoke condensate from starch pyrolysis had a different type of mutagen, which was active to TA100 without metabolic activation. The problem was raised of how to identify precursors of mutagens produced by pyrolysis of protein. Mutagenicity of tar produced by pyrolysis of various amino acids was studied. As shown in Table 2, tryptophan, serine and glutamic acid produced strong mutagenicity after pyrolysis (Nagao et al., 1977c). The process of isolating and identifying the mutagenic principle was started by using pyrolysates of amino acids and proteins, and cooked food. Mutagen formation in beef and beef extract during cooking was reported (Commoner et al., 1978).

This paper reviews our studies which have been carried out, as a collaborative effort, by the Cancer Institute, Shizuoka Pharmaceutical College, the University of Tokyo, the University of Tohoku, Keio University and the American Health Foundation.

Chemical structure and analysis of mutagens produced by pyrolysis

Several new mutagens isolated from pyrolysates of amino acids, proteins and food are summarized in Table 3. Two new mutagens, 3-amino-1,4-dimethyl-5H-pyrido[4,3-b]indole (Trp-P-1) and 3-amino-1-methyl-5H-pyrido[4,3-b]indole (Trp-P-2) were isolated from a basic fraction of pyrolysate of 330 g DL-tryptophan by a combination of column-chromatography techniques, such as those using aluminium oxide, CM-Sephadex and Sephadex LH-20 (Sugimura et al., 1977a). Chemical structure was elucidated by mass spectrum, nuclear magnetic resonance spectra and X-ray crystallography (Kosuge et al., 1978) and confirmed

TABLE 1

MUTAGENICITY OF SMOKE CONDENSATES OBTAINED BY PYROLYSIS OF BIOLOGICAL MATERIALS (Nagao et al., 1977a)

	Revertants/mg smoke condensates			
	TA98		TA100	
	+S9 mix	−S9 mix	+S9 mix	−S9 mix
Lysozyme	8311	0	2319	0
Histone	5012	0	1311	0
DNA	278	0	170	0
RNA	83	0	0	0
Starch	0	0	70	338
Vegetable oil	0	0	85	0

TABLE 2

MUTAGENICITY OF TAR FROM AMINO ACIDS AND AMINES ON S. typhimurium TA98 WITH S9 MIX Nagao et al., 1977c)

Amino Acid	Revertants/mg tar
Tryptophan	22000
Serine	18400
Glutamic acid	13800
Creatinine	10200
Creatine	8860
Ornithine	8290
Lysine	5250
Arginine	4950
Citrulline	1850
Threonine	3100
Alanine	2980
Cystine	2140
Glutamine	1600
Methionine	890
Cysteine	324
Tyrosine	199
Phenylalanine	148
Histidine	104
Asparagine	98
Valine	0.9

by chemical synthesis (Akimoto et al., 1977; Takeda et al., 1977). Two new mutagens, 2-amino-6-methyldipyrido[1,2-a:3′,2′-d]imidazole (Glu-P-1) and 2-aminodipyrido[1,2-a:3′,2′-d]imidazole (Glu-P-2) were isolated from a basic fraction of pyrolysate of 10 kg L-glutamic acid (Takeda et al., 1978; Yamamoto et al., 1978). 3,4-Dicyclo-pentenopyrido[3,2-a]carbazole (Lys-P-1) was isolated from pyrolysate of L-lysine (Wakabayashi et al., 1978). 2-Amino-5-phenylpyridine, which had previously been isolated as an antifungal substance from tar of D,L-phenylalanine (Kosuge and Zenda, 1976) was proven to be mutagenic (Sugimura et al., 1977a).

2-Amino-α-carboline (AαC) and 2-amino-3-methyl-α-carboline (MeAαC) were isolated from pyrolysate of soybean globulin by Yoshida et al. (1978). Trp-P-1 and Trp-P-2 were detected in pyrolysates of casein and gluten (Uyeta et al., 1979). Glu-P-2 was detected in a pyrolysate of casein (Yamaguchi et al., 1979).

Recently, 2 unique mutagens were isolated from sun-dried and broiled sardine (Kasai et al., 1979). One of them was also isolated from heated beef (Spingarn et al., 1980). These were 2-amino-3-methylimidazo[4,5-f]quinoline (IQ) and 2-amino-3,4-dimethylimidazo[4,5-f]quinoline (MeIQ) and they proved to be very strong mutagens toward S. typhimurium TA98 with S9 mix (Kasai et al., 1980a, b).

Trp-P-1 and Trp-P-2 were detected in sun-dried sardine broiled by ordinary cooking methods and estimated as 13.3 and 13.1 ng per g broiled sardine, respectively, by gas chromatography and mass spectrometry (Yamaizumi et al., 1980). Trp-P-1 was also detected in broiled beef (Yamaguchi et al., 1980b). Glu-P-2 was detected in sun-dried cuttlefish broiled in the ordinary way (Yamaguchi et al., 1980a).

TABLE 3
NEW MUTAGENS FROM PYROLYSATES

Chemical name (Abbreviation)	Chemical structure	Source	Specific mutagenicity revertants/μg TA98, +S9 mix	In vitro carcinogenicity	In vivo carcinogenicity	Ref.
3-Amino-1,4-dimethyl-5H-pyrido[4,3-b]indole (Trp-P-1)		Tryptophan pyrolysate	39000	+	+	Sugimura et al. (1977a)
3-Amino-1-methyl-5H-pyrido[4,3-b]indole (Trp-P-2)		Tryptophan pyrolysate	104000	+	+	Sugimura et al. (1977a)
2-Amino-6-methyldipyrido[1,2-a:3′,2′-d]imidazole (Glu-P-1)		Glutamic acid pyrolysate	49000	+	a	Yamamoto et al. (1978)
2-Amino-dipyrido[1,2-a:3′,2′-d]imidazole (Glu-P-2)		Glutamic acid pyrolysate	1900	—	a	Yamamoto et al. (1978)
2-Amino-5-phenylpyridine (Phe-P-1)		Phenylalanine pyrolysate	41	b	b	Sugimura et al. (1977a)
3,4-Cyclopentenopyrido[3,2-a]carbazole (Lys-P-1)		Lysine pyrolysate	86	a	b	Wakabayashi et al. (1978)
2-Amino-α-carboline (AαC)		Soybean globulin pyrolysate	300	a	a	Yoshida et al. (1978)
2-Amino-3-methyl-α-carboline (MeAαC)		Soybean globulin pyrolysate	200	a	a	Yoshida et al. (1978)

2-Amino-3-methylimidazo-[4,5-*f*]quinoline (IQ)	Broiled sardine	433000	a	a	Kasai et al. (1980a,b)
2-Amino-3,4-dimethyl-imidazo[4,50-*f*]quinoline (MeIQ)	Broiled sardine	661000	b	a	Kasai et al. (1980b)

[a] On going. [b] Not tested yet.

Biochemical and biological activities of mutagens produced by pyrolysis

Trp-P-1 and Trp-P-2 induced in vitro cell transformation of cryopreserved embryonic cells from hamsters (Takayama et al., 1977, 1978). Cells transformed by Trp-P-2 were subcultured and formed fibrosarcomas when inoculated into the cheek pouches of young hamsters (Takayama et al., 1979a). Trp-P-1 induced sarcomas in Syrian golden hamsters and Fischer rats by subcutaneous injection of 1.5 mg once a week for 20 weeks (Ishikawa et al., 1979). Trp-P-2 did not induce tumor in either hamsters or rats under the same experimental conditions (Ishikawa et al., 1979). Trp-P-1 and Trp-P-2 produced liver tumors in CDF_1 mice who were given a diet containing 0.02% of Trp-P-1 or Trp-P-2, and female mice were more susceptible to these carcinogens than males (Matsukura et al., 1981). Glu-P-1 also induced morphological transformation of cryopreserved embryonic cells from hamsters (Takayama et al., 1979b), but Glu-P-2 did not induce cell transformation.

Trp-P-2, Trp-P-1, Glu-P-1 and AαC induced sister-chromatid exchanges (SCEs) in human lymphoblastoid cells. (They are listed here in declining order of potency for SCE induction.) All these compounds required an in vitro metabolic activation system (S9 mix) for the induction of SCEs (Tohda et al., 1980).

Trp-P-1 and Trp-P-2 required P-448 systems of rat-liver microsome for their activation. The metabolic activation system was reconstructed with purified cytochrome P-448 or P-450, NADPH-cytochrome P-450 reductase, dilauroyl-L-3-phosphatidyl choline and NADPH. P-448, purified from rat liver pretreated with polychlorinated biphenyls (PCB), and P-448, purified from rat liver pretreated with 3-methyl cholanthrene both activated, very efficiently, Trp-P-1 and Trp-P-2. P-450, purified from rat liver pretreated with either PCB or phenobarbital, was less active in activation of Trp-P-1 and Trp-P-2 (Ishi et al., 1980). An active metabolite of Trp-P-2, activated by rat-liver microsomes, was bound to DNA and polyG, but not to polyA, polyC and polyU (Hashimoto et al., 1978).

Mutagenic activities of Trp-P-1, Trp-P-2, Glu-P-1 and AαC were eliminated by an action of peroxidase and H_2O_2 (Yamada et al., 1979). Horseradish peroxidase and myeloperoxidase were active on degradation of Trp-P-1 and lactoperoxidase was less active. Trp-P-2 was more unstable than Trp-P-1 and Glu-P-1 (Yamada et al., 1979). Trp-P-1, Trp-P-2 and Glu-P-1 were inactivated by 0.05 mM nitrite at acidic pH (Tsuda et al., 1980). Glu-P-1 was the most unstable, with a half-life less than 5 min at pH 1.6. These heterocyclic aromatic amines were deaminated by nitrite and the corresponding hydroxy derivatives were identified (Tsuda et al., 1980).

The formation of mutagens and carcinogens in the process of cooking food was thus elucidated. These carcinogens present in food might play an important role in the development of human cancer and this possibility should be taken into serious consideration. We have to study, more thoroughly, the mechanism of the formation of these mutagens and carcinogens by pyrolysis and the mechanism of inactivation of these mutagens and carcinogens. We should also increase our efforts to reduce the levels of these mutagens and carcinogens in our daily food.

Acknowledgements

This work was supported, in part, by grants from the Ministry of Education, Science and Culture, the Ministry of Health and Welfare, the Princess Takamatsu Cancer Research Fund and the Society for Promotion of Cancer Research.

References

Akimoto, H., A. Kawai, H. Nomura, M. Nagao, T. Kawachi and T. Sugimura (1977) Syntheses of potent mutagens in tryptophan pyrolysates, Chem. Lett., 1061—1064.

Ames, B.N., J. McCann and E. Yamasaki (1975) Methods for detecting carcinogens and mutagens with the Salmonella/mammalian-microsome mutagenicity test, Mutation Res., 31, 347—364.

Commoner, B., A.J. Vithayathil, P. Dolara, S. Nair, P. Madyastha and G.S. Cuca (1978) Formation of mutagens in beef and beef extract during cooking, Science, 201, 913—916.

Haenszel, W., and M. Kurihara (1968) Studies of Japanese migrants, I. Mortality from cancer and other diseases among Japanese in the United States, J. Natl. Cancer Inst., 40, 43—68.

Hashimoto, Y. K. Takeda, K. Shudo, T. Okamoto, T. Sugimura and T. Kosuge (1978) Rat liver microsome-mediated binding to DNA of 3-amino-1-methyl-5H-pyrido[4,3-b]indole, a potent mutagen isolated from tryptophan pyrolysate, Chem. -Biol. Interact., 23, 137—140.

Ishii, K., M. Ando, T. Kamataki, R. Kato and M. Nagao (1980) Metabolic activation of mutagenic tryptophan pyrolysis products (Trp-P-1 and Trp-P-2) by a purified cytochrome P-450-dependent monoxygenase system, Cancer Lett., 9, 271—276.

Ishikawa, T., S. Takayama, T. Kitagawa, T. Kawachi, M. Kinebuchi, N. Matsukura, E. Uchida and T. Sugimura (1979) In vivo experiments on tryptophan pyrolysis products, in: E.C. Miller et al. (Eds.), Naturally Occurring Carcinogens—Mutagens and Modulators of Carcinogenesis Jpn. Sci. Soc. Press., Tokyo/Univ. Park Press, Baltimore, pp. 159—167.

Kasai, H., S. Nishimura, M. Nagao, Y. Takahashi and T. Sugimura(1979) Fractionation of a mutagenic principle from broiled fish by high-pressure liquid chromatography, Cancer Lett., 7, 343—348.

Kasai, H., S. Nishimura, K. Wakabayashi, M. Nagao and T. Sugimura (1980a) Chemical Synthesis of 2-amino-3-methylimidazo[4,5-f]quinoline (IQ), a potent mutagen isolated from broiled fish, Proc. Jpn. Acad., 56B, 382—384.

Kasai, H., Z. Yamaizumi, K. Wakabayashi, M. Nagao, T. Sugimura, S. Yokoyama, T. Miyazawa, N.E. Spingarn, J.H. Weisburger and S. Nishimura (1980b) Potent novel mutagens produced by broiling fish under normal conditons, Proc. Jpn. Acad., 56, 278—283.

Kosuge, K., K. Tsuji, K. Wakabayashi, T. Okamoto, K. Shudo, Y. Iitaka, A. Itai, T. Sugimura, T. Kawachi, M. Nagao, T. Yahagi and Y. Seino (1978) Isolation and structure studies of mutagenic principles in amino acid pyrolysates, Chem. Pharm. Bull. 26, 611—619.

Kosuge, T., and H. Zenda (1976) Studies on active principles in dry distillation tars from plants and animals, Yuki Gosei Kagaku Kyokai Shi, 34, 612—624 (in Japanese).

McCann, J., E. Choi, E. Yamasaki and B.N. Ames (1975a) Detection of carcinogens as mutagens in the Salmonella/microsome test: Assay of 300 chemicals, Proc. Natl. Acad. Sci. (U.S.A.), 72, 5135—5139.

McCann, J., N.E. Spingarn, J. Kobori and B.N. Ames (1975b) Detection of carcinogens as mutagens: bacterial tester strains with R-factor plasmid, Proc. Natl. Acad. Sci. (U.S.A.), 72, 979—983.

Matsukura, N., T. Kawachi, K. Morino, H. Ohgaki, T. Sugimura and S. Takayama (1981) Carcinogenicity in mice of mutagenic compounds from a tryptophan pyrolysate, Science, in press.

Matsushima, T., T. Sugimura, M. Nagao, T. Yahagi, A. Shirai and M. Sawamura (1980) Factors modulating mutagenicity in microbial tests in: K.H. Norpoth et al. (Eds.), Short-Term Test System for Detecting Carcinogens, Springer, Berlin, pp. 273—285.

Nagao, M., M. Honda, Y. Seino, T. Yahagi, T. Kawachi and T. Sugimura (1977a) Mutagenicities of protein pyrolysates, Cancer Lett., 2, 335—340.

Nagao, M., M. Honda, Y. Seino, T. Yahagi and T. Sugimura (1977b) Mutagenicities of smoke condensates and the charred surface of fish and meat, Cancer Lett., 2, 221—226.

Nagao, M., T. Yahagi, T. Kawachi, Y. Seino, M. Honda, N. Matsukura, T. Sugimura, K. Wakabayashi, K. Tsuji and T. Kosuge (1977c) Mutagens in foods, and especially pyrolysis products of protein, in: D. Scott et al. (Eds.), Progress in Genetic Toxicology, Elsevier/North-Holland, Amsterdam, pp. 259—264.

Spingarn, N.E., H. Kasai, L.L. Vuolo, S. Nishimura, Z. Yamaizumi, T. Sugimura, T. Matsushima and J.H. Weisburger (1980) Formation of mutagens in cooked foods, III. Isolation of a potent mutagen from beef, Cancer Lett., 9, 177—183.

Sugimura, T., S. Sato, M. Nagao, T. Yahagi, T. Matsushima, Y. Seino, M. Takeuchi and T. Kawachi (1976) Overlapping of carcinogens and mutagens, in: P.N. Magee et al. (Eds.), Fundamentals in Cancer Prevention, Jpn. Sci. Soc. Press, Tokyo/Univ. Park Press, Baltimore, pp. 191—215.

Sugimura, T., T. Kawachi, M. Nagao, T. Yahagi, Y. Seino, T. Okamoto, K. Shudo, T. Kosuge, K. Tsuji, K. Wakabayashi, Y. Iitaka and A. Itai (1977a) Mutagenic principle(s) in tryptophan and phenylalanine pyrolysis products, Proc. Jpn. Acad., 53, 58—61.

Sugimura, T., M. Nagao, T. Kawachi, M. Honda, T. Yahagi, Y. Seino S. Sato, N. Matsukura, T. Matsushima, A. Shirai, M. Sawamura and H. Matsumoto (1977b) Mutagen—carcinogens in food, with special reference to highly mutagenic pyrolytic products in broiled foods, in: H.H. Hiatt et al. (Eds.), Origins of Human Cancer, Cold Spring Harbor Lab., Cold Spring Harbor, pp. 1561—1577.

Takayama, S., Y. Katoh, M. Tanaka, M. Nagao, K. Wakabayashi and T. Sugimura (1977) In vitro transformation of hamster embryo cells with tryptophan pyrolysis products, Proc. Jpn. Acad., 53B, 126—129.

Takayama, S., T. Hirakawa and T. Sugimura (1978) Malignant transformation in vitro by tryptophan pyrolysis products, Proc. Jpn, Acad., 54B, 418—422.

Takayama, S., T. Harakawa, M. Tanaka, Y. Katoh and T. Sugimura (1979a) Transformation and neoplastic development of hamster embryo cells after exposure to tryptophan pyrolysis products in tissue culture, in: E.C. Miller et al.(Eds.), Naturally Occurring Carcinogens—Mutagens and Modulators of Carcinogenesis, Jpn. Sci. Soc. Press, Tokyo/Univ. Park Press, Baltimore, pp. 151—157.

Takayama, S., T. Hirakawa, M. Tanaka, T. Kawachi and T. Sugimura (1979b) In vitro transformation of hamster embryo cells with a glutamic acid pyrolysis products, Toxicol. Lett., 4, 281—284.

Takeda, K., T. Ohta, K. Shudo, T. Okamoto, K. Tsuji and T. Kosuge (1977) Synthesis of a mutagenic principle isolated from tryptophan pyrolysate, Chem. Pharm. Bull., 25, 2145—2146.

Takeda, K., K. Shudo, T. Okamoto and T. Kosuge (1978) Synthesis of mutagenic principles isolated from L-glutamic acid pyrolysate, Chem. Pharm. Bull., 26, 2924—2925.

Tohda, H., A. Oikawa, T. Kawachi and T. Sugimura (1980) Induction of sister-chromatid exchanges by mutagens from amino acid and protein pyrolysate, Mutation Res., 77, 65—69.

Tsuda, M., Y. Takahashi, M. Nagao, T. Hirayama and T. Sugimura (1980) Inactivation of mutagens from pyrolysates of tryptophan and glutamic acid by nitrite in acidic solution, Mutation Res., 78, 331—339.

Uyeta, M., T. Kanada, M. Mazaki, S. Taue and S. Takahashi (1979) Assaying mutagenicity of food pyrolysis products using the Ames test, in: E.C. Miller et al. (Eds.), Naturally Occurring Carcinogen—Mutagens and Modulators of Carcinogenesis, Jpn. Sci. Soc. Press, Tokyo/Univ. Park Press, Baltimore, pp. 169—176.

Wakabayashi, K., K. Tsuji, T. Kosuge, K. Takeda, K. Yamaguchi, K. Shudo, T. Iitaka, T. Okamoto, T. Yahagi, M. Nagao and T. Sugimura (1978) Isolation and structure determination of a mutagenic substance in L-lysine pyrolysate, Proc. Jpn. Acad., 54B, 569—571.

Yamada, M., M. Tsuda, M. Nagao, M. Mori and T. Sugimura (1979) Degradation of mutagens from pyrolysates of tryptophan, glutamic acid and globulin by myeloperoxidase, Biochem. Biophys. Res. Commun., 90, 769—776.

Yamaguchi, K., H. Zenda, K. Shudo, T. Kosuge, T. Okamoto and T. Sugimura (1979) Presence of 2-aminodipyrido[1,2-a:3',2'-d]imidazole in casein pyrolysate, Gann, 70, 849—850.

Yamaguchi, K., K. Shudo, T. Okamoto, T. Sugimura and T. Kosuge (1980a) Presence of 2-aminodipyrido-[1,2-a:3', 2'-d]imidazole in broiled cuttlefish, Gann, 71, 743—744.

Yamaguchi, K., K. Shudo, T. Okamoto, T. Sugimura and T. Kosuge (1980b) Presence of 3-amino-1,4-dimethyl-5H-pyrido[4,3-b]indole in broiled beef, Gann, 71, 745—746.

Yamaizumi, Z., T. Shiomi, H. Kasai, S. Nishimura, Y. Takahashi, M. Nagao and T. Sugimura (1980) Detection of potent mutagens, Trp-P-1 and Trp-P-2, in broiled fish, Cancer Lett., 9, 75—83.

Yamamoto, K., K. Tsuji, T. Kosuge, T. Okamoto, K. Shudo, K. Takeda, Y. Iitaka, K. Yamaguchi, Y. Seino, T. Yahagi, M. Nagao and T. Sugimura (1978) Isolation and structure determination of mutagenic substances in L-glutamic and pyrolysate, Proc. Jpn, Acad., 54B, 248—250.

Yoshida, D., T. Matsumoto, R. Yoshimura and T. Matsuzaki (1978) Mutagenicity of amino-α-carbolines in pyrolysis products of soybean globulin, Biochem. Biophys. Res. Commun., 83, 915—920.

MUTAGENIC PESTICIDES

THE GENOTOXICITY OF BENOMYL

A. KAPPAS

Biology Department, Nuclear Research Center Democritus, Athens (Greece)

Introduction

In 1968 certain benzimidazole derivatives emerged as promising fungicides (Erwin et al., 1968), although it had been known long before that benzimidazole itself acted as growth inhibitor in several living systems such as yeasts and bacteria (Woolley, 1944). Prominent in the group of benzimidazole fungicides are benomyl, known also under the trade name of benlate (methyl-1-(butylcarbamoyl-2-benzimidazole carbamate), and its stable breakdown product MBC or carbedazin (methyl-2-benzimidazole carbamate) (Clemons and Sisler, 1969), which are shown in Table 1. Other fungicides of this group, important although less active, are thiabendazole or TBZ, 2-(4-thiazolyl)-benzimidazole (Allen and Gottlieb, 1970), which is also used as anthelminthic (Brown et al., 1961), and fuberidazole, 2,2-(2'-furyl)-benzimidazole.

Important also and of similar activity, although not resembling a benzimidazole, are the thiophanate fungicides such as methyl thiophanate shown in Table 1 (1,2-bis(3-methoxycarbonylthioureido)benzene). The thiophanates are considered to belong to the benzimidazole group of fungicides, because, as shown by Vonk and Kaars Sijpesteijn (1971), they are converted under natural conditions to benzimidazole compounds which are responsible for the fungitoxic effect.

Benomyl is highly toxic to a variety of fungi, particularly to Ascomycetes, whereas it is completely ineffective to Oomycetes (Delp and Klopping, 1969). It is a systemic fungicide so can enter plant cells and efficiently translocates to the growing parts of the plant thus protecting it from disease. Because of its high systemic activity, benomyl became of outstanding importance for agriculture.

Clemons and Sisler (1969) have shown that benomyl breaks down rapidly in aqueous solutions to form MBC which is stable and actually the toxic material in the plant where it is applied for disease control. Because of this, in this paper I shall refer to both benomyl and MBC as the same fungitoxic compound.

The first evidence of the toxicity of MBC to fungal conidia was the inhibition of germ-tube extension as shown by Clemons and Sisler (1971) in

TABLE 1
BENZIMIDAZOLE FUNGICIDES

Compound	Structure
Benzimidazole	
Methyl-2-benzimidazole carbamate (MBC)	
Methyl-1-(butylcarbamoyl)-2-benzimidazole carbamate (benomyl or benlate)	
2-(4'-Thiazolyl)-benzimidazole (thiabendazole or TBZ)	
Fuberidazole	
1,2-bis(3-Methoxycarbonylthioureido) benzene (methyl thiophanate)	

Neurospora crassa. The same investigators showed that MBC caused inhibition of DNA synthesis when conidia of *N. crassa* were exposed for 3—4 h to the chemical. DNA synthesis was strongly inhibited at 8 h. These first studies on the toxicity of benomyl led the authors to suggest that the primary action of MBC is on DNA synthesis or some closely related process such as cell division. But because inhibition of DNA synthesis became evident several hours after treatment this inhibition might be considered as a secondary effect.

Genetic effects in *Aspergillus nidulans*

Almost parallel to the investigations for the understanding of the mode of action of benomyl and other benzimidazole fungicides, studies on the genotoxicity of these chemicals were also conducted.

The focusing of the fungal nucleus and its division as a target site of the toxic effect of benomyl, and the existence on the other hand of the benzimidazole ring in its molecule, which can be regarded as a structural analogue of purine, attracted attention to the possible genetic effects of the compound. Thus, in a short report by Hastie (1971) and then in a more extensive study by Kappas et al. (1974) it was shown for the first time that benomyl induces genetic instability in the Ascomycete *Aspergillus nidulans*.

Material and methods

In our studies we used the system of the parasexual cycle of *A. nidulans* based upon the pioneer work of Pontecorvo et al. (1953). This system, I think, is an ideal test system for studying the effects of environmental chemicals on somatic recombination.

There are 3 steps in the parasexual cycle of *A. nidulans*. (1) The formation of heterokaryons containing different haploid nuclei. (2) Fusion of 2 nuclei and formation of a diploid nucleus. (3) The segregation of diploid nuclei either by crossing-over which leads to the formation of recombinant diploid nuclei or by non-disjunction which, through aneuploidy, leads to the formation of haploid or recombinant diploid nuclei.

We synthesized a stable heterozygous diploid strain, the genotype of which is shown in Fig. 1, from 2 haploid strains derived from stocks held in the Genetics Department of the University of Glasgow. One haploid produced yellow conidia (y), had a suppressor gene for the adenine requirement, was resistant to acriflavin and also had all the other markers shown in Fig. 1.

The other haploid produced white conidia (w), was resistant to acriflavin and required p-aminobenzoic acid, biotin, thiamine and ammonium (cnx). The synthesized diploid strain was prototrophic and produced green conidia.

Somatic recombination occurs spontaneously in the diploid strains of *A. nidulans* but at a very low frequency. We found that this frequency greatly increased when we grew the strain in medium containing benomyl. The segregation can easily be observed and scored as white and yellow sectors in the growing green colonies of the fungus.

With the Aspergillus test system we not only score for the occurrence of mitotic recombination by counting the number of colour sectors but we can distinguish from the phenotypes of the segregants at least 3 mechanisms by which such recombination might occur, i.e. non-disjunction, mitotic crossing-over and breakage and deletion (Kappas, 1978).

Thus, as we can see from Fig. 1, white mitotic cross-over diploids must be prototrophs, but non-disjunctionals must require thiamine and ammonium. Yellow mitotic cross-over diploids must require adenine whereas non-disjunctionals must be prototrophs. White haploids must always require thiamine and ammonium. Yellow haploids must require riboflavin or ammonium from chromosome 8 and maybe one or more of the requirements in the other chromosomes of the diploid strain.

Results and discussion

In different experiments we tried several concentrations of benomyl, which was dissolved in ethanol and added to the growth medium where blocks of MM covered with growing mycelium were transferred to form colonies.

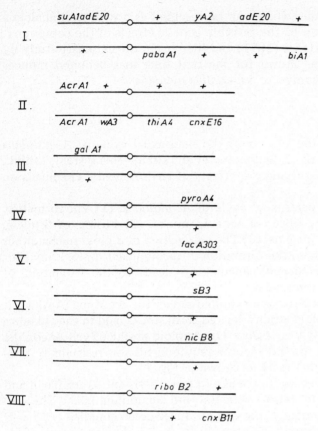

Fig. 1. Genotype of the heterozygous diploid strain of *Aspergillus nidulans*.

Table 2 shows the effect of benomyl on mitotic recombination in *A. nidulans*. Very low concentrations of benomyl, which had a toxic effect up to 50% measured as a reduction of colony diameter, greatly increased both haploid and diploid segregants. Analysis of the diploid segregants showed that there was no significant increase in mitotic cross-overs. Most of the diploid recombinants were due to non-disjunction, the mechanism that was also responsible for the haploid recombinants.

It is possible that some diploid recombinants scored as mitotic cross-overs were hemizygous for the respective chromosomal regions resulting from chromosome breaks and terminal deletions. This is illustrated in Fig. 2 for the class of yellow diploid recombinants requiring adenine. These recombinants could be the result of either mitotic crossing-over on chromosome I or breaks and deletions of segments of the homologous chromosome carrying the wild-type allele for yellow colour (Kappas, 1978).

To find out which of the two mechanisms was responsible for the formation of the *yad* 1st-order recombinants, a number of such recombinants was further analysed by determining the phenotypes of the spontaneous 2nd-order white haploid recombinants formed in the yellow colonies. If the *yad* recombinants were formed by mitotic crossing-over, then the white haploids would either

TABLE 2

RECOMBINOGENIC EFFECT OF BENOMYL ON THE DIPLOID STRAIN OF *Aspergillus nidulans*

Compound	Concentration (μM)	Toxicity[a]	Mitotic segregants[b] per 100 colonies		
			Haploid	Diploid	
				Non-dis-junctional	Mitotic cross-over[c]
Control	—	—	15	5	45
Benomyl	0.5	18	70	60	50
	1.0	32	138	77	47
	1.5	54	255	100	40

[a]Percentage reduction of colony diameter, measured 3 days after inoculation.
[b]White or yellow sectors, patches or very small spots in the colonies after 7 days of incubation.
[c]Includes those that could have been due to breakage-deletion.

require adenine or not, but if they were formed by breaks and deletions then none of them would require adenine (Fig. 3). Although with benomyl there was not a significant increase of diploid segregants considered as mitotic crossovers, because even untreated colonies showed a rather increased number compared with what is known for spontaneous segregation (Käfer, 1961; Roper, 1966), we tested a number of yellow *ad* diploid segregants (Table 3) and found that all these segregants were formed only by mitotic crossing-over and not by breaks and deletions.

Thus genetic analysis demonstrated that benomyl had a strong effect on non-disjunction in *A. nidulans*. These results were in accord with those of Bighami et al. (1977) who found that benomyl strongly induced non-disjunction in a different strain of *A. nidulans*.

Furthermore, the observation that with increasing concentration of benomyl

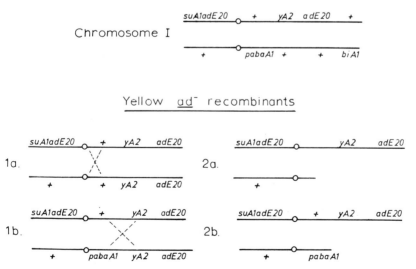

Fig. 2. Yellow diploid recombinants requiring adenine formed by either mitotic crossing-over (1a, 1b) or breakage—deletion (2a, 2b) in chromosome I.

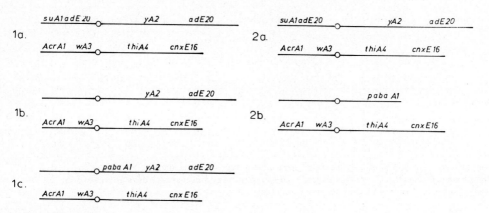

Fig. 3. White haploid 2nd-order recombinants from yad^- formed by mitotic crossing-over (1a, 1b, 1c) or by breakage and deletion (2a, 2b).

there was an increase in the number of sectors parallel to the increased inhibition of growth (Table 2) lends support to the hypothesis that the toxicity of benomyl is directly related to its effect on mitosis by causing non-disjunction.

This hypothesis was supported by investigations of Davidse (1975) on the toxicity of benomyl who found that the inhibition of growth of *A. nidulans* could be ascribed to an interference with mitosis by the fungicide. Davidse first suggested a mode of action of MBC similar to that of colchicine which is the formation of a complex between the chemical and a sub-unit of the microtubule proteins resulting in inhibition of spindle formation. By using cell-free extracts of mycelium of *A. nidulans* incubated with ^{14}C-labelled MBC, he deter-

TABLE 3

SPONTANEOUS SECOND-ORDER WHITE RECOMBINANTS FROM FIRST-ORDER YELLOW ad^- DIPLOIDS INDUCED BY BENOMYL IN *Aspergillus nidulans*

yad^- diploids		Second-order white haploids		
Treatment (μM)	Isolate No.	Total	ad^+	ad^-
Control	1	19	10	9
	2	26	14	12
	3	19	12	7
	4	22	10	12
	5	14	6	8
Benomyl 0.5	1	10	7	3
0.5	2	15	9	6
0.5	3	26	12	14
1.0	4	17	7	10
1.0	5	22	12	10
1.0	6	24	14	10
1.0	7	14	9	5
1.0	8	21	9	12

mined the binding of the chemical to fungal tubulin by the standard gel-filtration procedure. Then he purified the binding protein by subjecting it to SDS—polyacrylamide gel electrophoresis. By using as reference brain tubulin, he found that a protein was present with electrophoretic properties similar to those of brain tubulin which might be considered as the fungal tubulin, responsible for MBC binding. This finding evidently indicates that MBC acts as a spindle poison in fungi, thus causing non-disjunction.

Cytological investigations (Davidse, 1973) also gave evidence for the effect of MBC on abnormal cell division in *A. nidulans*.

Studies in other systems

Studies on the effects of benomyl or MBC in other systems include plants, mammals and microorganisms.

Richmond and Phillips (1975) had studied the effect of benomyl on cell division of *Botrytis cinerea* and onion root-tip cells. They found, by cytological examination, abnormalities in cell division similar to that caused by griseofulvin, an antifungal antibiotic which is known as an antimitotic agent binding to microtubules and thus preventing normal spindle formation and function (Roobol et al., 1977). Richmond and Phillips also found a rather small increase in chromosomal aberrations.

Similar effects of benomyl were observed cytologically in onion root tips by Boyle (1973) and also by Richmond and Pring (1977).

Regarding investigations on the genetic effects of benomyl on mammals, it was Styles and Garner (1974) who first studied these effects in rats and cultures of mammalian cells including human adult liver cells, HeLa cells, hamster kidney cells, etc. In rats, numerous mitotic abnormalities were found in bone marrow after oral administration or intraperitoneal injection. Following mitotic delay in cell culture, benomyl at a low level caused chromosome breakage and bridge formation. Metaphase arrest and micronuclei in cell cultures were observed also (Styles and Garner, 1974).

In dominant lethal studies in rats, Sherman et al. (1975) found that benomyl at the highest dietary level tested (0.25%) was ineffective.

Seiler (1976), by using the micro-nucleus test, found that MBC produces micronuclei in mouse bone-marrow cells through inhibition of mitosis.

Experiments in rats by Sheerman et al. (1975) showed that benomyl had no embryotoxic or teratogenic effect. Benomyl was also ineffective in mitotic gene conversion studies in *Saccharomyces cerevisiae* (Siebert et al., 1970).

Point mutations

Regarding point mutations in mutagenicity tests with bacteria, Seiler (1972) found that MBC induced low levels of base-pair substitutions. Considering the structural analogy between benzimidazole and the purine nucleus, it seems possible that some incorporation of benzimidazole instead of a purine base into DNA may occur. Forward mutations have also been reported in the fungus *Fusarium oxysporum* (Dassenoy and Meyer, 1973).

By using the fluctuation test, as described by Green et al. (1976), we

TABLE 4

MUTAGENICITY OF BENOMYL IN Aspergillus nidulans UT439 AND A uvs D53 DERIVATIVE UT517 IN THE FLUCTUATION TEST

Strain[a]	Benomyl concentration (μg/ml)	Number of experiments	Total number of tubes tested	Total number of tubes positive		Probability[b]
				bi	pyr	
UT439	0	5	250	5	5	
	0.25	5	250	3	4	
	0.30	5	250	6	6	
	0.40	5	250	4	5	
UT 517	0	5	250	4	4	
	0.25	5	250	70	80	<0.001
	0.30	5	250	90	85	<0.001
	0.40	5	250	100	110	<0.001

[a]Both strains require proline, p-aminobenzoic acid, biotin and pyridoxine.
[b]Significance by χ^2 test as described by Green et al. (1976).

(Kappas et al., 1976) found that benomyl mutates enterobacteria by the lex$^+$-dependent misrepair pathway and that the premutagenic lesion is susceptible to excision repair. We suggested that benomyl may be a new type of base-analogue mutagen that, when incorporated into DNA, is detected by UV endonuclease and provokes mutations not by misincorporation during normal DNA replication but by misrepair of gaps of newly synthesized DNA.

We then studied the effect of benomyl in 2 strains of *A. nidulans* differing in the ability to excise pyrimidine dimers (Kappas and Bridges, 1981), again using the fluctuation test. The results are shown in Table 4. Benomyl induced back mutations from both biotin and pyridoxine requirement in a dose-dependent manner in the presumed excision-deficient UT517 strain, whereas there was no detectable mutagenic effect in the repair-proficient parent strain UT439.

Conclusion

The main genotoxic effect of benomyl is that connected with its mechanism of action as a fungicide. It seems that it affects the spindle formation most probably by incorporation into the microtubule protein thus causing abnormal cell division by non-disjunction. This occurs not only in *A. nidulans* but also in mammalian cells. These findings, together with the mutagenic effect of the chemical, imply a potential genetic risk for the human population.

References

Allen, P.M., and D. Gottlieb (1970) Mechanism of action of the fungicide thiabendazole 2-(4'-thiazolyl)-benzimidazole, Appl. Microbiol., 20, 919—926.

Bignami, M., F. Aulicino, A. Velcich, A. Carere and G. Morpurgo (1977) Mutagenic and recombinogenic action of pesticides in *Aspergillus nidulans*, Mutation Res., 46, 395—402.

Boyle, W.S. (1973) Cytogenetic effects of benlate fungicide on *Allium cepa* and *Secale cereale*, J. Hered., 64, 49—50.

Brown, H.D., A.R. Matzuk, I.R. Ilvues, L.H. Peterson, S.A. Harris, L.H. Sarett, J.R. Edgerton, J.J. Yaktis,

W.C. Campbell and A.C. Cuckler (1961) Antiparasitic drugs, IV. 2-(4'-Thiazolyl)-benzimidazole, a new anthelmintic, J. Am. Chem. Soc., 83, 1764—1765.

Clemons, G.P., and H.D. Sisler (1969) Formation of a fungitoxic derivative from benlate, Phytopathology, 59, 705—706.

Clemons, G.P., and H.D. Sisler (1971) Localization of the site of action of a fungitoxic benomyl derivative, Pest. Biochem. Physiol., 1, 32—43.

Dassenoy, B., and J.A. Meyer (1973) Mutagenic effect of benomyl on *Fusarium oxysporum*, Mutation Res., 21, 119—120.

Davidse, L.C. (1973) Antimitotic activity of methyl benzimidazol-2-yl carbamate (MBC) in *Aspergillus nidulans*, Pest. Biochem. Physiol., 3, 317—325.

Davidse, L.C. (1975) Antimitotic activity of methyl benzimidazol-2-ylcarbamate in fungi and its binding to cellular protein, in: M. Borgers and M. de Brabander (Eds.), Microtubules and microtubule inhibitors, North-Holland, Amsterdam, pp. 483—495.

Delp, C.J., and H.L. Klopping (1969) Disease control with fungicide 1991, in: 1st Int. Congr. Plant Pathol., p. 44.

Erwin, D.C., H. Mee and J.J. Sims (1968) The systemic effect of 1-(butylcarbamoyl)-2-benzimidazole carbamic acid, methyl ester, on verticillium wilt of cotton, Phytopathology, 58, 528.

Green, M.H.L., W.J. Muriel and B.A. Bridges (1976) Use of a simplified fluctuation test to detect low levels of mutagens, Mutation Res., 38, 33—42.

Hastie, A.C. (1971) Benlate induced instability of Aspergillus diploids, Nature (London), 226, 771.

Kafer, E. (1961) The processes of spontaneous recombination in vegetative nuclei of *Aspergillus nidulans*, Genetics, 46, 1581—1609.

Kappas, A. (1978) On the mechanisms of induced somatic recombination by certain fungicides in *Aspergillus nidulans*, Mutation Res., 51, 189—197.

Kappas, A., and B.A. Bridges (1981) Induction of point mutations by benomyl in DNA repair-deficient *Apergillus nidulans*, Mutation Res., 91, 115—118.

Kappas, A., S.G. Georgopoulos and A.C. Hastie (1974) On the genetic activity of benzimidazole and thiophanate fungicides on diploid *Aspergillus nidulans*, Mutation Res., 26, 17—27.

Kappas, A., M.H.L. Green, B.A. Bridges, A.M. Rogers and W.J. Muriel (1976) Benomyl — a novel type of base analogue mutagen? Mutation Res., 40, 379—382.

Pontecorvo, G., J.A. Roper, L.M. Hemmons, K.D. MacDonald and A.W.J. Bufton (1953) The genetics of *Aspergillus nidulans*, Adv. Genet., 5, 141—238.

Richmond, A., and A. Phillips (1975) The effect of benomyl and carbendazin on mitosis in hyphae of *Botrytis cinerea*, Pers. ex. Fr. and roots of *Allium cepa* L., Pest. Biochem. Physiol., 5, 367—379.

Richmond, D.V., and R.J. Pring (1977) Some cytological effects of systemic fungicides on fungi and plants, Neth. J. Plant Pathol., 83, Suppl. 1, 403—410.

Roobol, A., K. Gull and I. Pogson (1977) Evidence that griseofulvin binds to a microtubule associated protein, FEBS Lett., 75, 149—153.

Roper, J.A. (1966) The parasexual cycle, in: G.C. Ainsworth and A.S. Sussman (Eds.), The Fungi, Vol. 2, Academic Press, New York, pp. 589—617.

Seiler, J.P. (1972) The mutagenicity of benzimidazole and benzimidazole derivatives, I. Forward and reverse mutations in *Salmonella typhimurium* caused by benzimidazole and some of its derivatives, Mutation Res., 15, 273—276.

Seiler, J.P. (1976) The mutagenicity of benzimidazole and benzimidazole derivatives, VI. Cytogenetic effects of benzimidazole derivatives in the bone marrow of the mouse and the Chinese hamster, Mutation Res., 40, 339—348.

Sherman, H., R. Culik and R.A. Jackson (1975) Reproduction, teratogenic, and mutagenic studies with benomyl, Toxicol. Appl. Pharmacol., 32, 305—315.

Siebert, D., F.K. Zimmerman and E. Lemperle (1970) Genetic effects of fungicides, Mutation Res., 10, 533—543.

Styles, J.A., and R. Garner (1974) Benzimidazolecarbamate methyl ester-evaluation of its effects in vivo and in vitro, Mutation Res., 26, 177—187.

Vonk, J.W., and A. Kaars Sijpesteijn, A. (1971) Methyl benzimidazol-2-ylcarbamate, the fungitoxic principle of thiophanate-methyl, Pest. Sci., 2, 160—164.

Woolley, D.W. (1944) Some biological effect produced by benzimidazole and their reversal by purines, J. Biol. Chem., 152, 225—232.

DOES DICHLORVOS CONSTITUTE A GENOTOXIC HAZARD?

CLAES RAMEL

Wallenberg Laboratory, Environmental Toxicology Unit, University of Stockholm (Sweden)

Introduction

The rapid development of short-term methods for the screening of mutagenic and carcinogenic compounds has not been followed by any corresponding improvement in risk evaluation. The result is that regulatory agencies all over the world have the same headache in applying the data for regulatory actions. These difficulties particularly pertain to widely used chemicals, which have been reported as suspected mutagens and carcinogens. In the risk-benefit evaluation of such chemicals, the benefit may be obvious but usually difficult to estimate, while it is often next to impossible even to make any qualified guess of the risk situation. There are some chemicals of this kind, which have been the notorious subject of endless discussions, statements and accusations for many years and in many countries. Two examples of such chemicals are saccharin and dichlorvos.

Saccharin constitutes a good example of all the problems involved in handling a suspected carcinogen. Its use as a food additive has made the problem particularly difficult. The battle in the United States between the Congress, the Food and Drug Administration (FDA), the industry, the scientists and the general public has indeed been confusing. It is highly understandable when a representative of the FDA stated: "The FDA will prepare the ground more carefully next time it takes action to regulate a weak but popular carcinogen" (Smith, 1980).

How frustrating a case like saccharin may appear, it fulfils at least the important function of providing some experience of how to handle cases like that. The course of events concerning saccharin has many counterparts for other chemicals, although the publicity may not have reached the same level elsewhere.

Dichlorvos or DDVP is one of those chemicals, which have been subjected to somewhat similar discussions although for about 3 times as long a period. The obvious principle difference between the saccharin and the dichlorvos case is the fact that dichlorvos is used as a pesticide and is therefore not subjected to food regulations such as the Delanian clause in the U.S.A. However, in reality the difference is not that pronounced considering the kind of exposure of people also to dichlorvos. The predominant use of dichlorvos is as resin strips in rooms against flies and mosquitos. There is thus a continuous release of insecticide vapour from those strips. It should therefore be emphasized that the human exposure to that insecticide is rather unique. Man is in fact exposed to the same treatment as the insect, which he fights! In other words one has to rely on the difference in sensitivity to dichlorvos between insects and man. It is a priori highly understandable that many people have felt uncomfortable in this situation and wondered how much confidence one really can have in this stated difference in toxicity between insects and man. Does the lower susceptibility of man for instance really apply to all situations and to all human individuals? The concern about possible health hazards by dichlorvos exposure took a more serious course when it was shown that dichlorvos can alkylate DNA and is mutagenic in bacteria. The discussions which followed initiated a new and intense testing activity by scientists both in the manufacturing company Shell and elsewhere. The result of this is that our knowledge of how dichlorvos acts chemically and biologically is exceptionally extensive. This fact in combination with its wide use and the unusual exposure of man has made dichlorvos somewhat of a key substance of how to deal with mutagenic chemicals from a regulatory point of view. This was the reason why ICPEMC, the International Commission for Protection against Environmental Mutagens and Carcinogens, took up dichlorvos for evaluation and formed for this purpose an ad hoc work group, which has recently finished its report (Ramel et al., 1980).

An evaluation of the genotoxic effects of dichlorvos has also recently been performed by IARC, the International Agency for Research on Cancer (IARC, 1979).

In this talk I will try to make a brief survey of the biochemical and biological characteristics of dichlorvos and what conclusions can be drawn as to possible genotoxic risks by dichlorvos.

Metabolism of dichlorvos

In order to understand and evaluate the genotoxic effects of dichlorvos it is necessary to look into the chemical and biological characteristics of the molecule and its metabolism. The metabolic fate of dichlorvos in vitro and in vivo has been subjected to a large number of investigations which have been summarized in different connections. A very good review has recently been published by Wright et al. (1979).

Dichlorvos, 2,2-dichlorovinyl dimethyl phosphate has two parts with special interest from the point of view of its biological activity — the phosphoryl centre with the dichlorovinyl group and the methyl groups. These two parts are involved in the two metabolic pathways of dichlorvos.

Phosphorylation and hydrolysis

The electrophilic phosphoryl centre can be attacked by nucleophiles. This leads to the formation of a dimethylphosphorous bond with the nucleophile and to dichloroacetaldehyde. The nucleophilic attack on the phosphoryl centre is the major metabolic pathway at least in higher organisms, and it is responsible for both the insecticidal action of dichlorvos and to its detoxification in mammals. In insects the phosphoryl part of dichlorvos binds to acetylcholinesterase, which it inhibits and this causes the acute toxic effect. Mammals are apparently protected from this kind of a reaction by other esterases, arylesterases. The arylesterases instead catalyse the reaction of dichlorvos with water, which results in the formation of dimethyl phosphate and dichloroacetaldehyde. The difference in toxicity between insects and mammals therefore seems to rest on the absence of arylesterase in insects and presence in mammals. It may be a pertinent question to what extent we can rely on the presence of the catalysing esterases in all human beings. This is particularly justified as it has been shown by Augustinsson (1961) that the presence of arylesterase is under genetic control in pigs and the amount can vary from zero upwards. It would be interesting to study the susceptibility to dichlorvos of those pigs which lack arylesterase. A comparable variation in arylesterase has never been reported in humans as far as I know. Nor has anybody reacted towards dichlorvos like a mosquito to my knowledge, which may be reassuring.

The phosphorylation reaction of dichlorvos, whatever nucleophile is involved, leads to the formation of dichloroacetaldehyde. Now the formation of dichloroacetaldehyde implies a problem from a genetic viewpoint. Dichloroacetaldehyde has been shown to induce base-pair substitutions in Salmonella (Löfroth, 1978) and it has also been reported by Fischer et al. (1977) to induce dominant lethal mutations in mice at high dosages (176 mg/kg). On the other hand Becker and Schöneich (personal communication) did not observe any increase of dominant lethals in another strain of mice.

The further breakdown of dichloroacetaldehyde presumably gives rise to dichloroethanol, which is excreted as a glucuronide conjugate. Dichloroethanol did not show any mutagenic effects in Salmonella (Löfroth, 1978).

The metabolic degradation of dichlorvos through this pathway with an esterase-catalysed hydrolysis goes rapidly in all mammalian tissues studied. It has in fact been difficult to study the kinetics of dichlorvos metabolism *in vivo* because of its rapid biotransformation. This also applies to the intermediate metabolite dichloroacetaldehyde, which is rapidly converted to dichloroethanol or possibly other metabolites. The actual behavior of dichloroacetaldehyde is, however, of particular interest as this metabolite constitutes the only mutagenicity problem when it comes to the hydrolysis of dichlorvos in the mammalian body. It is true that dichlorvos labelled at the chlorine atom with ^{36}Cl was not detected in the tissues of pig (Page et al., 1972), but the chlorine bond is not stable and the resolving power of such a labelling is questionable. Although the presence of dichloroacetaldehyde from dichlorvos no doubt is of short duration in the body, a mutagenic effect of it is possible, at least from a theoretical point of view.

Demethylation

The hydrolysis of dichlorvos has turned out to be the most important pathway in the mammalian body at least at low doses. It is, however, not this pathway which has caused most concern from a mutational point of view, but rather the other of the two pathways involving demethylation of dichlorvos.

That pathway obviously involves a risk for mutation through methylation of DNA. Dichlorvos has also been shown to methylate DNA in bacteria and mammalian cells in vitro (Lawley et al., 1974; Wennerberg and Löfroth, 1974). The methylating activity of dichlorvos resembles rather much methyl methanesulphonate (MMS) although it reacts 50 times slower than MMS with DNA of HeLa cells (Lawley et al., 1974). The alkylation property of dichlorvos seems to be responsible for the mutations that occur in bacteria (Wright et al., 1979). This is indicated by the fact that the difference in mutagenicity in bacteria between MMS and dichlorvos is fairly well in accordance with the corresponding difference in alkylating properties.

From the point of view of risk assessment it is, however, more interesting to know what really happens in the mammalian body in vivo — whether dichlorvos can alkylate DNA there. Earlier attempts by Wooder et al. (1977) and by Wennerberg (1973) were unsuccessful in identifying any alkylation of DNA in vivo in rodents. Although this at least suggested that no major alkylation of DNA by dichlorvos occurs in mammals such negative results are not satisfactory for any risk evaluation. Some alkylation of DNA should occur also in vivo and a measurement of the amount is important. Recently Segerbäck (1981) has succeeded to measure this alkylation of DNA in mice in vivo. He used ^{14}C-labelled dichlorvos and measured the degree of alkylation of N-7 guanine. This enabled him to make a risk estimate by comparing to MMS. The ratio N-7/O-6 alkylation of guanine is the same for dichlorvos and MMS. The relation between alkylation and mutation, which is known for MMS, can therefore be used to estimate the mutation frequency of dichlorvos by the alkylation that Segerbäck recorded. It is furthermore possible to calculate the radiation equivalent dose that is how many rads of irradiation that a certain dose of dichlorvos corresponds to. The highest acceptable daily intake of dichlorvos have been suggested by WHO (1975) to be 4 μg/kg/day. Segerbäck showed that a person receiving that amount daily over a year would run a risk equivalent to receiving one millirad. This risk is indeed negligible and it therefore seems that the alkylating property of dichlorvos does not imply an unacceptable genetic risk.

Mutagenicity and dichlorvos

The total genetic effect of dichlorvos should be reflected in the mutagenicity data with the compound itself, and there is indeed an extensive amount of data available in that respect.

If one examines the mutagenicity data of dichlorvos, there is an overwhelming amount of experimental results from bacteria showing a mutagenic potency of the compound. Also there is convincing evidence that fungi behave in a similar way (Table 1). Now dichlorvos is one of the few compounds where our knowledge of the metabolism and biotransformation of the compound enables us to get beyond such mutational data on microorganisms to judge the genetic risk of dichlorvos to man. The fact that different metabolic pathways are used in different organisms makes it important to use metabolically relevant test systems before extrapolating to man. The mutagenic effects in bacteria can be accounted for by alkylation of DNA (Wright et al., 1979), which evidently plays an insignificant role in mammals in vivo.

TABLE 1

MUTAGENICITY TEST RESULTS ON MICROORGANISMS

Test organism	Point mutation	Induced recombination	Non-disjunction	Ref.
Bacteria				
E. coli Sd-4-Forw.	+			Löfroth et al. (1969)
WP2 Rev.	+			Ashwood-Smith et al. (1972)
WP2 Rev.	−			Dean (1972a)
WP2 Rev.	+			Bridges et al. (1973)
WP2 Rev.	+			Bridges (1978)
K12 Forw.	+			Mohn (1973)
K12 Forw.	+			Voogd et al. (1972)
B Forw.				Wild (1973)
Salmonella typhimurium Rev.	+			Dyer and Hanna (1973) Carere et al. (1978)
Forw.	+			Voogd et al. (1972)
Citrobacter froundii Forw.	+			Voogd et al. (1972)
Enterobacter aerogenes Forw.	+			Voogd et al. (1972)
Klebsiella pneumoniae Forw.	+			Voogd et al. (1972)
Serratia marcescens Rev.	+			Dean (1972a)
Streptomyces coelicolor Forw.	+			Carere et al. (1978)
Fungi				
Saccharomyces cerevisiae		+		Dean et al. (1972)
		+		Fahrig (1973)
Neurospora crassa	−			Michalek and Brockmann (1969)
Aspergillus nidulans	+		+	

If we turn to mutagenicity studies in higher organisms, the picture becomes far more complicated (Table 2). In Drosophila tests with recessive lethals have been negative, except for an experiment reported by Hanna and Dyer (1973). They studied, however, the accumulation of recessive lethals during 30 generations. The actual number of mutational events cannot be calculated in such an experiment and it is difficult to draw any safe conclusions as to the mutagenicity of dichlorvos from the results, even if there was a significantly higher frequency of lethal after 30 generations in the dichlorvos treated series. Gupta and Singh (1974) studied salivary-gland chromosomes after feeding dichlorvos to larvae and they reported an increase of inversions and deletions, but apparently no translocations. Such a result is a priori unexpected and attempts to repeat that experiment by Sobels' group (personal communication) have been unsuccessful. Therefore Gupta and Singh's data cannot be accepted without further confirmation. When discussing Drosophila, I would also like to stress that a test involving an insect can hardly be the most appropriate for the screening of an insecticide. As an illustration of that I may mention that we performed a recessive lethal test on Drosophila with dichlorvos 10 years ago with negative result. A parallel series with an equimolar dose of EMS was, however, also negative! Such a result can therefore hardly be considered sensible — but rather as no result at all. Nevertheless, our results happened to reach the information section of Shell company in Sweden and they sent out an information sheet with an account of our results and the happy conclusion that

TABLE 2

MUTAGENICITY TEST RESULTS ON HIGHER ORGANISMS

Test organism	Point mutations	Chromosome aberrations	SCE	Other	Ref.
Drosophila					
	(—)				Jayasuriya and Ratnayake (1973)
	—				Kramers and Knaap (1978)
	—				Sobels and Todd (1979)
	(+)				Hanna and Dyer (1975)
		(+)			Gupta and Singh (1974)
Plants					
Allium		+			Sax and Sax (1968)
Hordeum		+			Bhan and Kaul (1975)
		+			Panda and Sharma (1980)
Mammals					
In vivo					
Mice dominant lethals		—			Buselmaier et al. (1972)
		—			Dean and Blair (1976)
		—			Dean and Thorpe (1972b)
		—			Epstein et al. (1972)
Mice cytogenetics		—			Dean and Thorpe (1972a)
Chinese hamster		—			Dean and Thorpe (1972a)
Mice, sperm abnormalities				+	Wyrobek and Bruce (1975)
Host-mediated assay					
Mice	—				Blair et al. (1975)
	—				Buselmaier et al. (1972)
	—				Dean et al. (1972)
In vitro					
Human lymphocytes		—			Bootsma et al. (1971) Dean (1972b)
Human fibroblasts			(+)		Nicholas et al. (1978)
		+			Ishidate (1976)
Chinese hamster V79	—				Wild (1975)
		+	+		Tezuka et al. (1980)

dichlorvos did not imply a genetic risk! They did withdraw their information sheet at our request, however.

Both in plants and in mammalian cell cultures have chromosome aberrations been observed after treatment with dichlorvos and furthermore an increase of SCE has been recorded in human and hamster cells. The lack of appropriate metabolism makes it, however, difficult to extrapolate such data to an intact mammalian body.

When it comes to whole mammals no mutational effects have been recorded from any tests. One possible exception is an increase of abnormal sperms in mice as reported by Wyrobek and Bruce (1975). The mechanism behind the formation of such defects is however not clear-cut and involves other than genetic effects.

In summing up the mutagenicity studies with dichlorvos it is clear that this pesticide is mutagenic in several test systems in vitro, but there is no unambiguously positive result in vivo from whole animals in spite of several experiments being performed. It must however in that connection be stressed that only host-mediated assays take into consideration point mutations, but the resolving power with that method is not sufficient to rely on negative result.

Cancer data

4 long term cancer studies have been reported with dichlorvos — 3 in rats and 1 in mice. One of the cancer studies on rats (Blair et al., 1976) was a 2-year inhalation study, which should be of particular relevance considering that man is exposed in the same way. However, that experiment suffers from some shortcomings, such as very low dosages (up to 5 μg/l), high mortality and high spontaneous tumour incidence. Therefore, the fact that no increase of tumours could be detected by dichlorvos does not carry any strong weight. The 3 other cancer studies were feeding studies with higher exposure to dichlorvos. None of them showed any significant increase of tumours either. However, in one study, performed by NCI on mice, some rare tumours were recorded in the treated series (National Cancer Institute, 1977). Also in NCI's study on rats a dose-related increase in malignant fibrous histiocytomas was suggested in males but in none of these experiments was it possible to conclude whether the tumours were connected with dichlorvos treatment.

Long-term studies on rats are performed in Japan with dichlorvos administered at 280 and 140 ppm in drinking water. A preliminary report close to the termination of the experiment does not indicate any increased tumour incidence (Interim Report by Dr. M. Enomoto).

The cancer data on dichlorvos is thus negative from a statistical point of view, but the observations of tumours in the 2 experiments by NCI makes it necessary to interpret the negative data with some caution.

Trichlorphone

When evaluating experimental results on dichlorvos there is another set of data which also has to be included — that is from the drug trichlorphone or metrifonate. That drug has been widely used against schistosomiasis and it has been shown to convert to dichlorvos both in vivo and in vitro (Dedek et al., 1969; Nordgren et al., 1978). Trichlorphone has been studied by Fischer et al. (1977) who also reported dominant lethality by dichloroacetaldehyde. They also found an increased dominant lethality in mice by trichlorphone and they suggested that this effect was caused by the formation of dichloroacetaldehyde via dichlorvos.

In a cancer study on rats Gibel et al. (1973) reported a significant increase of malignant and benign tumours after treatment with trichlorphone. Other cancer studies have also been performed and some carcinogenic effects have been suggested. However these experiments have not been reported in such a form that they can be evaluated properly. After analyses of the cancer studies with trichlorphone WHO/FAO (FAO, 1979) did not consider a carcinogenic effect of this drug established.

Does dichlorvos imply a genetic risk?

Now comes the interesting question — what do we do with this mixture of positive and negative experimental data? Is dichlorvos really mutagenic and carcinogenic? Because of the extensive investigations of the biochemistry and metabolism of dichlorvos we are in a much better position here than with other chemicals to classify the experimental data regarding its relevance to man. The fact that dichlorvos is mutagenic in bacteria must thus be judged against the background that this mutagenicity is at least essentially dependent on alkylation of DNA and such alkylation seems to be of negligible importance in intact mammals.

Instead we have another pathway operating in mammals — that is the hydrolysis of dichlorvos which is catalysed by certain esterases. Test results from systems without such esterases can therefore not safely be extrapolated to man — not even qualitatively. Mutagenic effects in such test systems as plants and cell cultures must therefore be interpreted with great caution.

It is obvious that one has to pay special attention to tests with whole animals and to tests of metabolites which we know are found in the mammalian body.

Tests on whole animals, which can be assumed to reflect what happens in the human body have been negative but one has to keep in mind that no satisfactory test on point mutations in mammals have been available. This is an unfortunate situation since there are reasons to believe that point mutations are induced at lower doses than chromosome aberrations. Vogel has for instance shown that this is generally the case in Drosophila with alkylating agents (Vogel and Natarajan, 1979).

Against these negative results one must put the positive mutagenicity and carcinogenicity reports on trichlorphone, which furnish at least indirect indications of a possible genotoxic effect of dichlorvos.

When it comes to the metabolites of dichlorvos, the problem concentrates on dichloroacetaldehyde, which is a mutagen in bacteria. In mammals the dominant lethal data also suggest mutagenic effect but further data are badly needed. Even if dichloroacetaldehyde is a mutagen in mammals, however, the formation of that compound in the mammalian body does not necessarily mean a genotoxic risk, because of its very short lifelength. At present I do not consider that question settled and there is a need for further examination of the role of dichloroacetaldehyde as a metabolite from dichlorvos.

In conclusion, taken all experimental evidence on dichlorvos together, one can make the following evaluation. The effect of dichlorvos on microorganisms and in vitro is probably not of particular relevance to man. The most relevant test results have been negative, but there still are some experimental gaps to be filled. A weak mutagenic effect can therefore not be absolutely excluded. However, if such an effect occurs, it must be of very small magnitude and there is in my opinion no immediate need for a regulary action against the use of dichlorvos. The dichlorvos issue, however, has to be kept under continuous attention until supplementary data have been furnished.

The fact that dichlorvos still requires further consideration in spite of the unusually massive investigations performed illustrates the enormous difficulties to regulate the use of such biologically active compounds. The amount of mutagenic effects which can be tolerated with a widely used compound like dichlorvos is indeed small and therefore suspected mutagenic effects must be checked in greatest detail. If the remaining uncertainties about the genotoxic effects of dichlorvos are not solved, it may be advisable to eventually restrict that use of it which implies wide exposure to humans.

References

Ashwood-Smith, M.J., J. Trevino and R. Ring (1972) Mutagenicity of dichlorvos, Nature (London), 240, 418—420.
Augustinsson, K.-B., and B. Olsson (1961) Genetic control of arylesterase in the pig, Hereditas, 47, 1—22.
Bhan, A.K., and B.L. Kaul (1975) Cytogenetic activity of dichlorvos in barley, Indian J. Exp. Biol., 13, 403—405.
Blair, D., E.C. Hoadley and D.H. Hutson (1975) The distribution of dichlorvos in the tissues of mammals after inhalation and intravenous administration of dichlorvos, Toxicol. Appl. Pharmacol., 31, 243—253.
Blair, D., K.M. Dix, P.F. Hunt, E. Thorpe, D.E. Stevenson and A.I.T. Walker (1976) Dichlorvos — a two year inhalation carcinogenesis study in rats, Arch. Toxicol., 35, 281—294.
Bootsma, D., H. Herring, W. Kleijar, L. Budke, L.O.A. de Jong and F. Berends (1971) The effect of dichlorvos on human cells in tissue culture, Med.-Biol. Lab. RVO—TNO, MBL, 5.
Bridges, B.A. (1978) On the detection of volatile liquid mutagens with bacteria: experiments with dichlorvos and epichlorhydrin, Mutation Res., 54, 367—371.
Bridges, B.A., R.P. Mottershead, M.H.L. Green and W.J.H. Gray (1973) Mutagenicity of dichlorvos and methyl methanesulphonate for *Escherichia coli* WP 2 and some derivatives in DNA repair, Mutation Res., 19, 295—303.
Buselmaier, W., G. Röhrborn and P. Propping (1972) Mutagenitätsuntersuchungen mit Pestiziden im host mediated assay und mit dem dominanten Letaltest an der Maus, Biol. Zbl., 91, 311—325.
Carere, A., V.A. Ortali, G. Cardamone and G. Morpurgo (1978) Mutagenicity of dichlorvos and other structurally related pesticides in Salmonella and Streptomyces, Chem.-Biol. Interact., 22, 297—308.
Dean, B.J. (1972a) The mutagenic effects of organophosphorus pesticides on microorganisms, Arch. Toxicol., 30, 67—74.
Dean, B.J. (1972b) The effect of dichlorvos on cultured human lymphocytes, Arch. Toxicol., 30, 75—85.
Dean, B.J., and D. Blair (1976) Dominant lethal assay in female mice after oral dosing with dichlorvos or exposure to atmospheres containing dichlorvos, Mutation Res., 40, 67—72.
Dean, B.J., and E. Thorpe (1972a) Cytogenetic studies with dichlorvos in mice and Chinese hamsters, Arch. Toxicol., 30, 39—49.
Dean, B.J., and E. Thorpe (1972b) Studies with dichlorvos vapour in dominant lethal mutation tests on mice, Arch. Toxicol., 30, 51—59.
Dean, B.J., S.M.A. Doak and H. Funnel (1972) Genetic studies with dichlorvos in the host-mediated assay and in liquid medium using *Saccharomyces cerevisiae*, Arch. Toxicol., 30, 61—66.
Dedek, W., H. Koch, G. Uhlenhut and F. Bröse (1969) Zur Kenntnis der Umsetzung von ^{3}H-Trichlorphon zu DDVP, Z. Naturforsch., 24b, 663—664.
Dyer, K.F., and P.J. Hanna (1973) Comparative mutagenic activity and toxicity of triethylphosphate and dichlorvos in bacteria and Drosophila, Mutation Res., 21, 175—177.
Fahrig, R. (1973) Nachweis einer genetischen Wirkung von Organophosphor-Insektiziden, Naturwissenschaften, 60, 50—51.
Fischer, G.W., P. Schneider and H. Scheufler (1977) Zur Mutagenität von Dichloroacetaldehyde und 2,2-Dichlor-1,1-Dihydroxyäthanphosphonsäuremethylester, möglichen Metaboliten des phosphororganischen Pesticides Trichlorphon, Chem.-Biol. Interact., 19, 205—213.
Food and Agriculture Organization of the United Nations (FAO) (1979) Pesticide residues in food, 1978, FAO Plant Production and Protection Paper, 15 Rev. 27.
Gibel, W., Kh. Lohs, G.P. Wildner, D. Ziebarth and R. Stieglitz (1973) Über die Kanzerogene, hämatotoxische und hepatotoxische Wirkung pestizider organischer Phosphorverbindungen, Arch. Geschwulstforsch., 41, 311—328.
Gupta, A.K., and J. Singh (1974) Dichlorvos (DDVP) induced breaks in the salivary gland chromosomes of *Drosophila melanogaster*, Curr. Sci., 43, 661—662.
Hanna, P.J., and K.F. Dyer (1975) Mutagenicity of organophosphorus compounds in bacteria and Drosophila, Mutation Res., 28, 405—420.
IARC (1979) Monographs on the Evaluation of the Carcinogenic Risk of Chemicals to Humans, Some Halogenated Hydrocarbons, 20, 97—127.

Ishidate, M. (1976) Annu. Rep. of the Cancer Research, Ministry of Health and Welfare (Jpn.), 750.
Jayasuriya, V.U., and W.E. Ratnayake (1973) Screening of some pesticides on *Drosophila melanogaster* for toxic and genetic effects, Drosophila Inform. Serv., 50, 184—186.
Kramers, P.G., and A.G.A.C. Knaap (1978) Absence of a mutagenic effect after feeding dichlorvos to larvae of *Drosophila melanogaster*. Mutation Res., 57, 103—105.
Lawley, P.D., S.A. Shah and D.J. Orr (1974) Methylation of nucleic acids by 2,2-dichlorovinyl dimethyl phosphate (dichlorvos, DDVP), Chem.-Biol. Interact., 8, 171—182.
Löfroth, G. (1978) The mutagenicity of dichloroacetaldehyde, Z. Naturforsch., 33c, 783—785.
Löfroth, G., C. Kim and S. Hussain (1969) Alkylation property of 2,2-dichlorovinyl phosphate: a disregarded hazard, EMS News lett., 2, 21—26.
Michalek, S.M., and H.E. Brockman (1969) A test for mutagenicity of Shell "No-Pest Strip Insecticide" in *Neurospora crassa*, Neurospora Newslett., 14, 8.
Mohn, G. (1973) 5-Methyltryptophan resistance mutations in *Escherichia coli* K-12, Mutagenic activity of monofunctional alkylating agents including organophosphorus insecticides, Mutation Res., 20, 7—15.
National Cancer Institute (1977) Bioassay of dichlorvos for possible carcinogenicity (Carcinogenesis Technical Report Series, No. 10), DHEW Publication No. (NIH 77—810), Washington DC, U.S. Government Printing Office.
Nicholas, A.H., M. Vienne and H. van den Berghe (1978) Sister chromatid exchange frequencies in cultured human cells exposed to an organophosphorus insecticide: dichlorvos, Toxicol. Lett., 2, 271—275.
Nordgren, I., M. Bergström, B. Holmstedt and M. Sandoz (1978) Transformation and action of Metrifonate, Arch. Toxicol., 41, 31—41.
Page, A.C., J.E. Loeffler, H.R. Hendrickson, C.K. Huston and D.M. de Vries (1972) Metabolic fate of dichlorvos in swine, Arch. Toxicol., 30, 19—27.
Panda, B.B., and C.B.S.R. Sharma (1980) Cytogenetic hazards from agricultural chemicals, 3. Monitoring the cytogenetic activity of trichlorphon and dichlorvos in *Hordeum vulgare*. Mutation Res., 78, 341—345.
Ramel, C., J. Drake and T. Sugimura (1980), ICPEMC publication No. 5, An evaluation of the genetic toxicity of dichlorvos, Mutation Res., 76 (1980) 297—309.
Sax, K., and H.J. Sax (1968) Possible mutagenic hazards of some food additives, beverages and insecticides, Jpn. J. Genet., 43, 89—94.
Segerbäck, D. (1981) Estimation of genetic risks of alkylating agents, V. Methylation of DNA in the mouse by DDVP (2,2-dichlorovinyl-dimethyl phosphate), Hereditas, 94, 73—76.
Smith, R.J. (1980) Latest saccharin tests kill FDA proposal, Science, 208, 154—156.
Sobels, F.H., and N.K. Todd (1979) Absence of a mutagenic effect of dichlorvos in *Drosophila melanogaster*, Mutation Res., 67, 89—92.
Tezuka, H., N. Ando, R. Suzuki, M. Terahata, M. Moriya and Y. Shirasu (1980) Sister-chromatid exchanges and chromosomal aberrations in cultured Chinese hamster cells treated with pesticides positive in microbial reversion assays, Mutation Res., 78 (1980) 177—191.
Vogel, E., and A.T. Natarajan (1979) The relation between reaction kinetics and mutagenic action of monofunctional alkylating agents in higher eukaryotic systems, I. Recessive lethal mutations and translocations in Drosophila. Mutation Res., 62, 51—100.
Voogd, C.E., J.J.J.A.A. Jacobs and J.J. van der Stel (1972) On the mutagenic action of dichlorvos, Mutation Res., 16, 413—416.
Wennerberg, R. (1973) The methylation products after treatment with dichlorvos and dimethyl sulphate in vivo, Ph.D. Diss., University of Stockholm.
Wennerberg, R., and G. Löfroth (1974) Formation of 7-methylguanine by dichlorvos in bacteria and mice, Chem.-Biol. Interact., 8, 339—348.
Wild, D. (1973) Chemical induction of streptomycin-resistant mutations in *Escherichia coli*, Dose and mutagenic effects of dichlorvos and methyl methanesulfonate, Mutation Res., 19, 33—41.
Wild, D. (1975) Mutagenicity studies on organophosphorus insecticides, Mutation Res., 32, 133—150.
Wooder, M.F., A.S. Wright and L.J. King (1977) In vivo alkylation studies with dichlorvos at practical use concentrations, Chem.-Biol. Interact., 19, 25—46.
World Health Organization (WHO) (1975) Evaluations of some pesticide residues in food, WHO Pestic. Residue Ser., No. 4, 539.
Wright, A.S., D.H. Hutson and M.F. Wooder (1979) The chemical and biochemical reactivity of dichlorvos, Arch. Toxikol., 42, 1—18.
Wyrobek, A.J., and W.R. Bruce (1975) Chemical induction of sperm abnormalities in mice, Proc. Natl. Acad. Sci. (U.S.A.), 72, 4425—4429.

THE MUTAGENICITY OF THE FUNGICIDE THIRAM

M. ZDZIENICKA, M. ZIELENSKA, M. HRYNIEWICZ, M. TROJANOWSKA, M. ZALEJSKA and T. SZYMCZYK

Department of Biochemistry, Warsaw Medical School, Banacha 1, 02-097 Warsaw (Poland)

Introduction

On the basis of knowledge derived from epidemiological studies it is generally accepted that environmental factors are of major importance in a high proportion (80—90%) of human cancers (Higginson, 1980). It is a sad reflection on experimental work that the majority of human carcinogens has been identified by population studies.

Mutagenicity screening systems of the type developed by Ames have provided an important, additional technique in the search for carcinogenic substances. Increasing numbers of chemical carcinogens are being identified by using short-term tests along with epidemiological data and experiments on animals, so that the environment contamination by carcinogens may be clarified during the next few years.

Considerable attention is now being devoted to pesticides whose widespread use has led to measurable pollution of our environment. Since the patenting by Tisdale and Williams (1934) of the use of derivatives of dithiocarbamic acid as fungicides, numerous papers have been published on biological studies and practical applications of this group of compounds. The dithiocarbamates are widely used on seeds, vegetables etc., and as industrial fungicides on paper, plastics and textiles. They are also used in the rubber and plastics industries to increase the rate of vulcanization by interaction with sulphur.

Thiram (bis-dimethylthiocarbamoyl-disulfide) is an important member of the dimethyldithiocarbamate class of fungicides. The available data are insufficient for evaluation of the carcinogenicity of this compound (IARC, 1976; Innes et al., 1969). Thiram was tested for mutagenicity in the Salmonella assay and was found to be positive, without metabolic activation, against TA100, TA1530 and TA1534 strains (Shirasu et al., 1977). Recently, Hedenstedt et al. (1979) simultaneously with Zdzienicka et al. (1979) reported on positive results with thiram in the Salmonella-microsome assay both directly and after metabolic activation.

Now, we have determined the ability of thiram to induce mutations in a few short-term tests. Different metabolic activation systems derived from various mammalian and plant organisms were also investigated.

Ames test

We have shown (Zdzienicka et al., 1979) that thiram induces mutations without metabolic activation only in TA1535 and TA100 strains of *S. typhimurium*. Thiram, at 100 µg/plate, increased the mutation frequency in these strains about 3-fold. The presence of S9 liver homogenate decreased the mutagenic activity of thiram, thus indicating detoxication of thiram by the S9 fraction. In the absence of metabolic activation, thiram did not induce mutations in strains TA1538 and TA98, but the number of revertants was raised about 2—3-fold in the presence of the S9 fraction from Aroclor-treated rats. Thus, this type of mutagenic activity could appear only after biotransformation of thiram.

Repair test

The use of DNA-repair mutants in the mutation test often provides an additional source of information on the mode of action of the compound tested or its metabolites, and also supporting evidence for mutagenic effects per se.

The repair test was performed as described by Ames et al. (1973). Table 1 shows that thiram produced an appreciable difference in zones of killing in TA1538 ($uvrB^-$) and TA1978 (uvr^+) strains of *S. typhimurium* only in the absence of metabolic activation.

According to Ames et al. (1973), if a tested compound is more toxic to TA1538 than to TA1978, its action can be due to a covalent reaction with DNA. The results obtained suggest that thiram can be covalently bound to DNA in the absence of a metabolic activation system whereas, in the presence of fraction S9 from Aroclor-treated rats, thiram is a simple intercalating agent.

Inductest

In bacteria, there exists a readily detectable reaction to DNA alteration that proceeds via a pre-existing or induced (SOS) pathway (Witkin, 1976). Lysogenic induction as well as mutagenesis results from a complex series of reactions,

TABLE 1

ZONES OF KILLING PRODUCED BY THIRAM IN TA1538 ($uvrB^-$) AND TA1978 (uvr^+) STRAINS OF *Salmonella typhimurium* IN THE PRESENCE OF A METABOLIC ACTIVATION SYSTEM

Compound tested	µg/plate	Diameter of zone of killing (mm)[a]			
		TA1538 ($uvrB^-$)	TA1978 (uvr^+)	TA1538 ($uvrB^-$)	TA1978 (uvr^+)
		−S9		+S9	
Thiram	50	13 ± 1	9 ± 1	9 ± 1	8 ± 1
Thiram	100	15 ± 1	10 ± 1	10 ± 1	9 ± 1
Mitomycin C	1	35 ± 2	15 ± 1	NT	NT
2-AF	50	NT	NT	15 ± 1	8 ± 1
Crystal violet	10	22 ± 1	21 ± 1	21 ± 1	19 ± 1

[a]Mean ±SD from 6 plates. NT, not tested.

usually triggered by damage to DNA. Since the original suggestion by Lwoff (1953), prophage induction has been used in screening for potential chemical carcinogens (Heinemann, 1971). An improved test was described recently by Moreau et al. (1976). There are some indications that the list of agents detected by inductests and by mutagenicity tests overlap only partially. An inductest may detect damage to DNA that is non-mutagenic toward tester bacteria, or when potentially mutagenic DNA damage is lethal to the commonly used tester bacteria (Benedict et al., 1977; McCann et al., 1975).

To evaluate the ability of thiram to damage DNA in other bacterial tests we performed induction or prophage λ in liquid medium as described by Moreau et al. (1976) for the inductest III. The results were negative. Thiram did not increase significantly induction of prophage λ either in absence or presence of the metabolic activation system (Fig. 1). This suggests that thiram does not induce an SOS repair system in bacteria.

Test with *Aspergillus nidulans*

The principal advantage of this simple and most rapid method using an eukaryotic organism is the great variety of genetic damages that can be analysed. The method allows to test, along with the induction of point mutations, different genetic events such as mitotic crossing-over and mitotic non-disjunction which cannot be detected in prokaryotes.

To determine the effect of thiram on induction of gene mutation in the eukaryotic organism, the *A. nidulans* test described by Morpurgo et al. (1979) and by Bignami et al. (1974, 1977) was used. Mitotic crossing-over and non-disjunction were tested in the P strain of *A. nidulans* by determination of the frequency of sectoring colonies by the method described by Fratello et al.

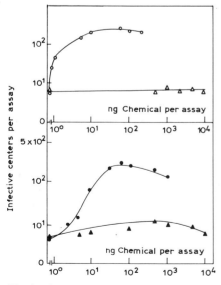

Fig. 1. The effect of thiram on prophage λ induction. Upper panel, prophage λ induction by mitomycin C(○) and thiram (△) in absence of S9. Lower panel, prophage λ induction by aflatoxin B_1 (●) and thiram (▲) in presence of S9.

(1960). Thiram induced point mutations in the plate and liquid tests only in the absence of the metabolic activation system. Thiram at 0.5—2 µg/plate raised about 6-fold the mutation frequency in the plate test (Table 2), and at 0.01—2.5 µg/ml about 4.5-fold in the liquid test (Table 3). In all these experiments, about 50% of the A. nidulans conidia survived at the highest doses of thiram.

The presence of the enzymatic activation system abolished the mutagenic activity of thiram in this test (Table 4).

Preliminary results indicate that thiram does not induce in A. nidulans either mitotic crossing-over or mitotic non-disjunction.

TABLE 2

MUTAGENIC ACTIVITY OF THIRAM IN THE Aspergillus nidulans SYSTEM su-meth (PLATE TEST)

Amount of thiram (µg/plate)	Survival (%)	Number of mutants per plate $\bar{x} \pm SE^a$	Number of mutants per 10^6 surviving conidia
0	100	2 ± 0.70	0.86
0.5	78.5	9 ± 0	4.9
1	67.4	8.5 ± 1.84	5.3
2	60	6.5 ± 1.25	4.6
MMS/positive control (1 µl) plate	52	40.2 ± 1.8	25.6

[a]The inoculum was 3×10^6 conidia/plate; $\bar{x}$, mean of 4 plates; SE, standard error.

TABLE 3

MUTAGENIC ACTIVITY OF THIRAM IN THE Aspergillus nidulans SYSTEM su-meth (LIQUID TEST, 1 h)

Amount of thiram (µg/ml)	Survival (%)	Number of mutants per plate $\bar{x} \pm SE^a$	Number of mutants per 10^6 surviving conidia
0	100	1.75 ± 0.25	0.7
0.01	78.3	6 ± 1.35	3.1
0.1	65.1	6 ± 0.70	3.1
1	50.9	4 ± 0.70	3.2
2.5	47.6	3.25 ± 0.94	2.75

[a]The inoculum was 3×10^6 conidia/plate; $\bar{x}$, mean of 4 plates; SE, standard error.

TABLE 4

MUTAGENIC ACTIVITY OF THIRAM IN THE Aspergillus nidulans SYSTEM su-meth IN THE PRESENCE OF S9 MIX (LIQUID TEST)

Amount of thiram (µg/ml)	Survival (%)	Number of mutants per plate $\bar{x} \pm SE^a$	Number of mutants per 10^6 surviving conidia
0	100	4 ± 0.91	2.3
0.01	81	3 ± 0.40	2.1
0.1	57.6	2 ± 0	2.0
1	44	1.75 ± 0.25	2.3

[a]The inoculum was 3×10^6 conidia/plate; $\bar{x}$, mean of 4 plates; SE, standard error.

Sperm abnormalities assay

The sperm morphology assay described by Wyrobek and Bruce (1975) is a simple method for assessing the effects of a variety of chemical and physical agents on induction of mutations in vivo. This short-term assay of possibly dangerous chemicals provides a new approach to monitoring health hazards. Studies with this assay have shown that mutagenic treatment during spermatogenesis leads to an increased frequency of sperm with abnormally shaped heads.

We applied this simple mammalian assay to measure the damage to spermatogenic cells in mice caused by thiram treatment in vivo. Table 5 presents preliminary results performed according to the method described by Wyrobek and Bruce (1975). Measurement of sperm abnormalities were made at 1 and 5 weeks after injections of chemicals. The control values for sperm abnormalities for animals treated only with oil were in the range of 0.7—1.2%, with a mean of 0.9%. The mean for sperm abnormalities was raised by B(a)P to 11.8% and by thiram to 19.4% after 5 weeks. The proportions of all forms of specific abnormalities were not changed in either case. Thiram interfered with normal differentiation of the germ cells, and the treatment produced a significant increase of abnormal sperm. This observation need not necessarily be interpreted as due to induced mutations. It is possible that thiram treatment produced some somatic changes in the animal which raised the level of abnormal sperm. As long as understanding of the mechanism of induction of sperm abnormalities is incomplete, interpretation of our results should be made with caution, and further studies are needed to elucidate the mechanism of action of thiram in mammals.

TABLE 5

MEANS AND SAMPLE SIZE FOR THE PERCENTAGE OF ABNORMAL SPERM F1 (CFW × C57BL)

Treatment	Dose (mg/kg b.w.)	1 Week	5 Weeks
Control	0	0.9 (5)	0.9 (5)
B/a/P[a]	100	0.8 (3)	11.8 (4)
Thiram[b]	100	1.1 (5)	19.4 (5)
Thiram	50	0.8 (4)	1.6 (4)

[a] B(a)P, 5 consecutive daily intraperitoneal injections.
[b] Thiram, single intraperitoneal injections.

The metabolic activation of thiram

Better knowledge of the metabolism of thiram is of importance for further evaluation of the risk involved in the application of this pesticide.

Numerous studies have established that, to become cytotoxic, mutagenic or carcinogenic, most chemicals have to be metabolically activated. The first stage in the metabolism of many compounds, both carcinogens and non-carcinogens, involves the mono-oxygenases that are present in the cells of most tissues in man and animals, and particularly in liver cells. The enzyme levels responsible for the biotransformation are raised or induced in animals treated with one of a large number of organic compounds, including carcinogens, and the choice of inducer can greatly modify the metabolic capacity and specificity of the resulting S9 liver homogenate preparation. Moreover, we can expect the existence of significant species differences.

Thiram requires metabolic activation before biological potency can be expressed in TA1538 and TA98 strains of *S. typhimurium* (Zdzienicka et al., 1979). The ability of thiram to cause mutations in these strains only in the presence of liver homogenate indicates that thiram is a promutagen converted by the rat-liver's metabolic action to the ultimate mutagen. The mutagenic activity of thiram was observed in the experiments with liver microsomes from Aroclor- and phenobarbital-treated rats, and also with microsomes from the liver of untreated mice (Fig. 2).

The pre-treatment of rats with a microsomal enzyme inducer had a dramatic influence on the mutagenic activity of thiram. The results show that the S9 fraction from untreated rats was not effective in catalysing conversion of thiram into mutagenic form(s). Negative results were also obtained by using human-liver microsomes, and when the activation system was lacking NADP. Thiram itself did not act as inducer of enzymes involved in its metabolic activation Fig. 2).

Metabolic activation of chemicals into mutagens is not confined to the mammalian assay system, but is a general phenomenon observed in both plant and animal systems (Plewa and Gentile, 1976). To determine whether thiram could be metabolized by some plants into mutagenic derivatives we applied the Ames procedure to evaluate biotransformation of thiram by plant extracts. The data summarized in Table 6 indicate that extracts derived from different plants did not convert thiram to mutagenic form(s) active toward any of the strains of *S. typhimurium*.

An extract of wheat sprouts had an anti-mutagenic activity. It decreased the

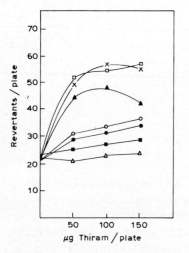

Fig. 2. The effect of different S9 fractions on mutagenic activity of thiram in *S. typhimurium* strain TA1538. S9 fractions from rat liver (○) not induced and induced by: (●) thiram; (□) Aroclor 1254; (■) Aroclor 1254 but without $NADP^+$ regenerating system; (▲) phenobarbital; (×) S9 fraction from mice liver; (△) S9 fraction from human liver.

TABLE 6

THE EFFECT OF S30 EXTRACT FROM VARIOUS PLANTS ON MUTAGENIC ACTIVITY OF THIRAM (100 μg/PLATE) IN THE *Salmonella typhimurium* TESTER STRAINS

Strain	Number of revertants/plate					
	−Thiram	+Thiram				
		−S30	+S30			
			Wheat sprouts	Carrot	Potato	Maize
TA1535	20	72	21	73	71	45
TA100	160	440	156	432	458	356
TA1538	25	27	26	25	26	25
TA98	40	44	41	45	42	42

TABLE 7

THE EFFECT OF S30 EXTRACT FROM VARIOUS PLANTS ON THE MUTAGENIC ACTIVITY OF THIRAM (100 μg/plate) IN TA1538 AND TA98 STRAINS OF *Salmonella typhimurium* IN THE PRESENCE OF S9 MIX FROM AROCLOR-TREATED RATS

Strain	Number of revertants/plate					
	−Thiram −S9	+Thiram +S9				
		−S30	+S30			
			Wheat sprouts	Carrot	Potato	Maize
TA1538	24	34	23	52	56	38
TA98	41	88	40	86	85	61

number of revertants induced by thiram to the control level (spontaneous reversion) in the TA1535 and TA100 strains. The anti-mutagenic activity of the wheat sprout extract was also observed in strains TA1538 and TA98 in the presence of S9 fraction of rat liver (Table 7). The biotransformation of thiram to ultimate mutagen depended on the source of the metabolic activation system. This finding indicates that the appropriateness of various metabolic activation systems in vitro should be experimentally established, and not simply assumed.

Our results lead to the conclusion that thiram is a mutagen, but it is not yet possible to extrapolate these results directly for the evaluation of the risk for man because no short-term assay for mutagenicity is completely successful in identifying carcinogens.

Acknowledgements

We are grateful to Dr. A. Carere who gave M. Zielenska the opportunity to perform the *A. nidulans* tests in his laboratory.

The investigation was sponsored by the Polish Academy of Sciences, project 09.7.2.1.13, and supported by funds provided in part by the International Cancer Research Data Bank Programme of the International Cancer Institute, the National Institute of Health (U.S.A.) under contract No. 1-00-65341 (International Cancer Research Technology Transfer, ICRETT) and the International Union Against Cancer.

References

Ames, B.N., F.D. Lee and W.E. Durston (1973) An improved bacterial test system for the detection and classification of mutagens and carcinogens, Proc. Natl. Acad. Sci. (U.S.A.), 70, 782.

Benedict, W.F., M.S. Baker, L. Haroun, E. Choi and B.N. Ames (1977) Mutagenicity of cancer chemotherapeutic agents in the Salmonella microsome test, Cancer Res., 37, 2209.

Bignami, M., G. Morpurgo, R. Pagliani, A. Carere, G. Conti and G. Di Giuseppe (1974) Non-disjunction and crossing-over induced by pharmaceutical drugs in *Aspergillus nidulans*, Mutation Res., 26, 159.

Bignami, M., F. Aulicino, A. Veicich, A. Carere and G. Morpurgo (1977) Mutagenic and recombinogenic action of pesticides in *Aspergillus nidulans*, Mutation Res., 46, 395.

Fratello, B., G. Morpurgo and G. Sermonti (1960) Induced somatic segregation in *A. nidulans*, Genetica, 45, 785.

Hedenstedt, A., U. Rannug, C. Ramel and C.A. Wachtmeister (1979) Mutagenicity and metabolism studies on 12 thiram and dithiocarbamate compounds used as accelerators in the Swedish Rubber Industry, Mutation Res., 68, 313.

Higginson, J.C.S. Muir (1980) Détermination de l'importance des facteurs environnementaux dans le cancer humain, rôle de l'épidémiologie, Bull. Cancer (Paris), 64, 365.

Heinemann, B. (1971) Prophage induction in lysogenic bacteria as a method of detecting potential mutagenic, carcinogenic, carcinostatic and teratogenic agents, in: A. Hollaender (Ed.), Chemical mutagens: Principles and Methods for Their Detection, Vol. 1, Plenum, New York, pp. 235—266.

Innes, J.M., B.M. Ulland, M.G. Valerio, L. Petrucelli, L. Fishbein, E.R. Hart, A.J. Pallotsa, R.B. Bates, H.L. Falk, J.J. Gart, M. Klein, I. Mitchell and J.J. Peters (1969) Bioassay of pesticides and industrial chemicals for tumorogenicity in mice; A preliminary note, J. Natl. Cancer Inst., 42, 1101.

International Agency for Research on Cancer (IARC) (1976) Monographs on the Evaluation of Carcinogenic Risk of the Chemicals to Man, Vol. 12, pp. 91—226.

Lwoff, A. (1953) Lysogeny, Bacteriol. Rev., 17, 269.

McCann, J., N.E. Spingarn, J. Kobori and B.N. Ames (1975) Detection of carcinogens as mutagens: Bacterial tester strains with R factor plasmids, Proc. Natl. Acad. Sci. (U.S.A.), 72, 979.

Moreau, P., A. Bailone and R. Devoret (1976) Prophage induction in *Escherichia coli* K12 envA uvr B: A highly sensitive test for potential carcinogens, Proc. Natl. Acad. Sci. (U.S.A.), 73, 3700.

Morpurgo, G., D. Bellincampi and C. Gualandi (1979) *Aspergillus nidulans* — Mutazione genitica, crossing-over mitotico e non disgiunzione mitotica, in: E. Magni (Ed.), Mutagenesi Ambientale Metodiche di Analisi, Vol. 1, Test in vitro, Consiglio Nazionale delle Ricerche, Roma, p. 169.

Plewa, M.J., and G. M. Gentile (1976) The mutagenicity of atrazine: a maize-microbe bioassay, Mutation Res., 38, 287.

Shirasu, Y., M. Moriya, K. Kato, F. Lienard, H. Tezuka, S. Teramoto and K. Kada (1977) Mutagenicity screening on pesticides and modification products: A basis of carcinogenic evaluation, in: H.H. Hiatt, J.D. Watson and J.A. Winsten (Eds.), Origins of Human Cancer, Book A, Cold Spring Harbor Laboratory, 267.

Tisdale, W.H., and I. Williams (1934) To E.I. DuPont de Nemours and Co., Disinfectant and fungicide, U.S.P. 1 972 961.

Witkin, E.M. (1976) Ultraviolet mutagenesis and inducible DNA repair *Escherichia coli*, Bacteriol. Rev., 40, 869.

Wyrobek, A.J., and W.R. Bruce (1975) Chemical induction of sperm abnormalities in mice, Proc. Natl. Acad. Sci. (U.S.A.), 72, 4425.

Zdzienicka, M., M. Zielenska, B. Tudek and T. Szymczyk (1979) Mutagenic activity of thiram in Ames tester strains of *Salmonella typhimurium*, Mutation Res., 68, 9.

COMPARISON OF THE MUTAGENIC ACTIVITY OF PESTICIDES in vitro IN VARIOUS SHORT-TERM ASSAYS

A. CARERE and G. MORPURGO

Istituto Superiore di Sanita, Viale R. Elena 299, Rome 00161, and Orto Botanico, University of Rome (Italy)

Introduction

In the general problems of environmental chemical mutagenesis and carcinogenesis, pesticides occupy a special position not only for their extensive use and widespread existence in the environment but also because they are chemicals very reactive from the biological point of view. Their potentially harmful effects apply to different extents to the entire human population. In fact, while a small fraction of people — the spraying men in agriculture or the workers in the pesticide industry — is exposed to high concentrations of these substances, the entire population is daily exposed to minute amounts of pesticides (or their metabolites or degradation products) in foods, households, etc.; the mutagenic and/or carcinogenic potential of pesticides may be small, but the dimensions of the population exposed to them could render the risk significant.

Thanks to the results of long- and short-term bioassays conducted in several laboratories during the last two decades, considerable information has now accumulated for tentative estimation of the potential risk of these substances.

The pesticides currently used for agricultural or domestic use are hundreds. Among them several have turned out to be unequivocally mutagenic and/or carcinogenic; others are suspected. Therefore a revision by the Health Authorities, of their use has become urgent.

In this paper we shall review part of the results obtained in a 5-year pesticide test program conducted by the Italian Public Health Institute in Rome, under the Environmental Research Programme of the Commission of the European Communities.

The most important objectives of the E.E.C. projects were the following. (1) To set up a battery of short-term assays in vitro to reveal different kinds of genetic damage. (2) To use this battery for comparative mutational studies on pesticides as pure compounds belonging to different chemical classes. (3) To try to relate the genetic activity with the chemical structure. (4) To devise an easy method to study the possible effects of plant metabolism on the genetic activity of pesticides.

Before presenting and discussing some of the results obtained we shall describe briefly the genetic systems used in our laboratory.

Short-term tests in vitro

The following mutation, recombination and DNA-repair tests have been used in vitro.

Prokaryotic tests

The well-known back-mutation genetic system developed by Ames et al. (1973) in *Salmonella typhimurium* was initially used with strains TA1535, TA1536, TA1537 and TA1538; since 1978 the TA98 and TA100 strains containing the R plasmid have also been used. For the metabolic activation system we used rat-liver microsomal fractions extracted from male rats of the Wistar race pre-treated with sodiumphenobarbital, or Sprague—Dawley male rats pre-treated with Aroclor 1254. The procedures followed were those described by Ames et al. (1975). Spot and plate tests were routinely used, and in rare cases a liquid test was also used.

During the last 2 years we have sometimes also used a forward-mutation assay, based on induction of 8-azaguanine resistance in *S. typhimurium*, slightly modified by Bignami and Crebelli (1979) with respect to the protocols proposed by Shopek et al. (1978). Only in one case (Benigni et al. 1979b) have we also used repair tests and selection of induced his^+ revertants according to the procedure described by Ames et al. (1973, 1975).

In parallel with the Ames test we have used a second bacterial test set up in the filamentous bacterium *Streptomyces coelicolor* A3 (Carere et al., 1978a, b; Bignami et al., 1980). It is a forward-mutation genetic system based on induction of resistants to low levels of streptomycin in the streptomycin-sensitive strain *hisA*1. The test was calibrated with direct or indirect alkylating agents to which it is particularly sensitive. The major limitation of the *S. coelicolor* genetic system is the small number of pro-mutagens and mutagens of the frame-shift type that can be detected. With this system, spot, plate and liquid tests were used.

Eukaryotic tests

The eukaryotic microorganism used in our laboratory was *Aspergillus nidulans*, an ascomycete widely exploited in genetic research. Its normal condition is haploid but stable diploid strains can easily be obtained, useful in the study of genetic events such as mitotic crossing-over and non-disjunction. With this eukaryotic microorganism the induction of point mutations was studied in the haploid strain 35 (*pabaA1, anA1, yA2, methG1, nicA2, nicB8*). The following 2 forward-mutation systems were used: induction of 8-azaguanine resistance, a completely recessive single-locus mutation very likely caused by the loss of a permease (Bignami et al., 1977), and induction of methionine suppression which involves at least 5 genes (Lilly, 1965).

Spot plate and liquid tests were used. Moreover, because *A. nidulans* is extremely resistant to variations in pH, the mutagenicity experiments can be performed with the medium at 3 different pHs (4.5, 7 and 8). The 2 forward-mutation systems are sensitive to various mutagens of the base-substitution type. It is not yet known whether they are also sensitive to frameshift mutagens. Another important limitation is the difficulty of detecting indirect mutagens by using the classic microsomal activation system. Recently, by using the so-called "growth-mediated assay", 6 pro-mutagens (dimethylnitrosamine, diethylnitrosamine, nitrosomorpholine, dimethylhydrazine, procarbazine and cyclophosphamide) were found positive in *A. nidulans*.

Somatic segregation (crossing-over and non-disjunction) was studied in the diploid strain P, whose genetic concentration regarding the first chromosome is shown in Fig. 1.

Because all the markers are recessive (Fig. 1) this heterozygous diploid strain is prototroph. It produces light-green conidia (because of the incomplete dominance of the *y* allele) and it is sensitive to *p*-fluorophenylalanine (FPA). On selective media supplemented with FPA and all the required nutrients the colonies can grow only if a mitotic crossing-over occurs between the centromere and the *fpaA1* locus or if the chromosome bringing the *fpaA1* allele becomes homozygous by non-disjunction. The 2 possibilities can easily be distinguished because in the case of mitotic crossing-over the *fpa*-resistant colonies will be pale-green ($yA2^+/yA2^-$) whereas in the second case colonies will be yellow ($yA2^+/yA2^+$). Moreover, the *fpa* cross-overs will not require nutrients if the crossing-over occurs in the region between *fpaA1* and *anA1* but they will require only aneurine if the crossing-over occurs between the centromere and *anA1*. The non-disjunctional yellow colonies resistant to FPA will require aneurine and paba. The spontaneous frequency of mitotic crossing-over is $1-3 \times 10^{-4}$ in a region whose meiotic length is about 40 morgans, while that of non-disjunction is about 1×10^{-5}. The analysis of somatic segregation was performed with both selective and non-selective methods. The selective technique, described by Bignami et al. (1974, 1977), consists in selecting, by addition of FPA directly to the plates, the homozygosity for the *fpa* marker, and is suitable only for the analysis of mitotic crossing-over; for this genetic event the selective method gives rapid results and may be performed with the spot-test technique. However, because it is possible that, even if only rarely, this method could produce false positive results because the selective action is exercised by the FPA that may itself induce non-disjunction, the analysis of this genetic event was performed with a non-selective method set up by Fratello et al. (1960). This method consists in scoring the increase of sectors on the surface of colonies grown on complete medium, devoid of FPA and supplemented with the chemical under test. With the P strain the segregants, non-disjunctional in the first chromosome, must be yellow ($yA2^-/yA2^-$) or dark green ($yA2^+/yA2^+$), depending on which of the 2 chromosomes of the diploid strain is in homozygosis or in hemizygosis.

Yellow sectors can derive either by crossing-over or by non-disjunction with or without production of haploid sectors; therefore, a genetic analysis of these sectors is necessary for a discrimination between cross-overs and non-disjunctions. Sectors derived by a crossing-over will require, at the maximum, paba. The non-disjunctional sectors will require aneurine and paba and will be FPA-resistant. Haploid sectors may require, in addition to aneurine and paba, other

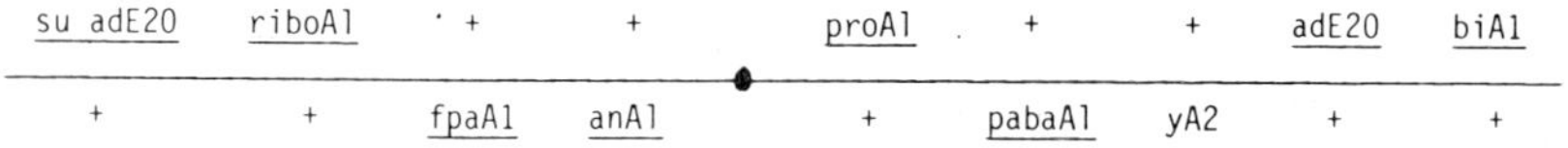

su adE20	riboA1	+	+		proA1	+	+	adE20	biA1
+	+	fpaA1	anA1		+	pabaA1	yA2	+	+

Fig. 1. Genotype of the strain P of *Aspergillus nidulans*.

nutrients, on the basis of the casual segregation of the other chromosomes. The spontaneous frequency of the colonies or yellow sectors (starting from the center of colonies) is about 3×10^{-3} for the crossing-over and 1×10^{-5} for non-disjunction and haploidization.

With the unselective method, plate and liquid tests were used for the analysis of non-disjunction. The first method gives only qualitative results because we treat with the chemical under test not single cells but whole populations of unknown dimensions. The liquid test is more laborious than the plate test but has the advantage that non-disjunction is induced on single conidia, thus permitting an exact quantitativeness and the construction of dose—response curves; moreover it can be applied either to quiescent or germinating conidia and allow us to try to relate the genetic activity with the physiological condition of the cell; this possibility permits one to have an idea about the target of the chemical under test in the induction of non-disjunction, i.e. DNA or mitotic spindle.

Technical details of such procedures for the analysis of non-disjunction were recently described by Morpurgo et al. (1979). By using such techniques they found that non-disjunction in *A. nidulans* can be caused by at least 3 classes of chemical: those, such as MMS, acting at DNA level; those, such as benomyl and FPA, acting at the mitotic spindle level; and those, such as amphotericin B and pimaricin, acting at the cell-membrane level. The major problem to be solved, a problem of great relevance for the environmental mutagenesis, is the extrapolation to higher organisms of the data obtained in *A. nidulans*, especially for the non-disjunctional agents acting not at the DNA level but on other targets such as mitotic spindles or membranes.

In *A. nidulans* a method to test the appearance of lethal recessives was set up by Morpurgo et al. (1978) which can be very useful because it permits detection not only of point mutations but also of any other type of genetic damage (deletions, translocations, etc.); this method, however, has so far been used with only few chemical agents.

Unscheduled DNA synthesis (UDS)

Since 1978 we have developed and used a DNA-damage repair test at cellular level, i.e. the induction of unscheduled DNA synthesis (UDS). The method set up in our laboratory (Benigni et al., 1979b) is essentially that described by San and Stich (1975) with only minor modifications. The radioactive incorporation of [^{3}H]thymidine was detected with an autoradiographic technique, by counting the number of grains in the nuclei of treated cells. This method was applied with EUE cells, a line of epithelial-like cells established in vitro from skin and muscle explants of human embryo. The experimental system was calibrated with standard physical and chemical agents which gave positive (UV, MNNG, MMS) and negative (ethydium bromide) results according to reports in the literature. Recently the UDS was also analyzed by the liquid-scintillation counting technique which seems to be much less sensitive than the autoradiographic technique (Benigni and Dogliotti, personal communication).

Results and discussion

We started our E.E.C. project by testing, with the 3 microorganisms (Salmonella, Streptomyces and Aspergillus), 12 pesticides of different use and chemical class (Table 1). The technique initially used was the "Spot test", applied with paper triangles of known dimensions embedded with the suspension of the substance to be tested. As an initial pre-screening to be used with compounds of unknown mutagenicity and especially for qualitative purposes, this technique has the advantage of permitting space—time gradients in the range of a few petri dishes (Bignami et al., 1974). In Table 2 the pesticides that had a positive effect in at least one genetic system are shown (Bignami et al., 1977; Carere et al., 1978). Aminotriazole(amitrole), a herbicide known to be carcinogenic in mice and rats (IARC, Vol. 7, 1974), which is banned in Italy, was positive in *S. coelicolor* and weakly positive in *A. nidulans* in the induction of both mitotic crossing-over and non-disjunction (confirmed by the non-selective method). The fungicide benomyl was positive only as inducer of non-disjunction. This compound was then intensively investigated in our laboratory by quantitative techniques; it is the most powerful non-disjunctional agent, acting at very low concentrations (2—4 µg/ml) and its target seems to be the mitotic spindle. Benomyl had already been reported as non-disjunctional by Hastie (1971) and by Kappas et al. (1974).

The 2 structurally related fungicides, captafol and captan, turned out to be positive as inducers of point mutations and mitotic crossing-over in *A. nidulans*; captan was also positive in the Ames test (strain TA1535) with and without S9. Captan, for which an abundant literature is available demonstrating unequivocally its genotoxic potential, is reported to be carcinogenic in mice by the NCI (Techn. Rep. No. 15, 1977) and is one of the pesticides for which we think a revision of its use is urgent, especially as to residues in the food chain.

The herbicide picloram induced gene mutations in the *S. coelicolor* system but it was completely negative in all other tests. This herbicide, in its liquid preparation known as tordon (triisopropanolamine salt of the picolinic acid), was mutagenic in *S. coelicolor*. Picloram was reported by the NCI to be carcinogenic in rats (Techn. Rep. No. 23, 1978).

Finally, dichlorvos, an insecticide banned in Italy for domestic use, was positive in the induction of point mutations in *S. coelicolor* and *A. nidulans*, in which it also induced mitotic crossing-over and non-disjunction.

By using in parallel the *S. typhimurium*, *S. coelicolor* and *A. nidulans* genetic systems we then performed mutational studies on 7 organophosphorus compounds and 5 carbamates (Table 3) in order to try to relate their genetic activity with their chemical structure. Table 4 shows the results published by Morpurgo et al. (1977) and by Carere et al. (1978). The chemical structures of the 7 organophosphates chosen are similar in one part of their molecules, i.e. the phosphoric ester moiety, but different in the other part. Dichlorvos is characterized by the presence of a vinylidene chloride group attached to the phosphoric part. Trichlorfon is known for its spontaneous conversion to dichlorvos. The other 5 organophosphates are devoid of the chlorinated group. Dichlorvos induced point mutations in all 3 microorganisms; as already reported, it also induced somatic segregation. Trichlorfon, which was weakly mutagenic in Salmonella (only in the liquid test), and *S. coelicolor*, in *A. nidulans* was

TABLE 1

PESTICIDES TESTED IN SALMONELLA, STREPTOMYCES AND ASPERGILLUS[a]

Common name	Systematic name	Use	Common name	Systematic name	Use
Aminotriazole	3-amino-1,2,4-triazole	herbicide	Dinobuton	2-sec-butyl-4,6-dinitro-phenyl isopropyl carbonate	acaricide and fungicide
Benomyl	methyl-N-1(butyl-carbamoyl)-2-benzimidazole carbamate	fungicide	Dodine	dodecyl guanidine acetate	fungicide
Captafol	N-(tetrachloroethylthio)-tetrahydrophthalimide	fungicide	Ioxynil	4-hydroxy-3,5-diiodo-benzonitrile	herbicide
Captan	N-(trichloromethylthio)-tetrahydrophthalimide	fungicide	Mecoprop	(±)-2-(4-chloro-2-methylphenoxy)propionic acid	herbicide
Dalapon-Na	Na-2,2-dichloropropionate	herbicide	Neburon	1-butyl-3(3,4-dichlorophenyl)-1-methylurea	herbicide
Dichlorvos	2,2-dichloro-vinyl-dimethyl phosphate	insecticide	Picloram	4-amino-3,5,6,trichloro-picolinic acid	herbicide

[a]Bignami et al. (1977); Carere et al. (1978a).

TABLE 2

MUTAGENIC AND RECOMBINOGENIC ACTIVITY OF PESTICIDES TESTED WITH A SPOT-TEST TECHNIQUE[a]

	S. typhimurium his+ revert.	S. coelicolor strept. resist.	A. nidulans Point mutation	A. nidulans Mitotic segregation c.o.	A. nidulans Mitotic segregation n-d
Aminotriazole	—	+	—	(+)	(+)
Benomyl	—	—	—	—	+++
Captafol	—	—	+	+	—
Captan	+(TA1535)	—	+	+	—
Dichlorvos	(+) (TA1535)	+	+	+	+
Picloram	—	+	—	—	—

[a]Bignami et al. (1977); Carere et al. (1978a).

TABLE 3

ORGANOPHOSPHORUS ESTERS AND CARBAMATES ASSAYED IN SALMONELLA, STREPTOMYCES AND ASPERGILLUS[a]

Organophosphorus esters	Systematic name	Carbamates	Systematic name
Azinphos methyl	S-(3,4-dihydro-4-oxobenzene-d-1,2,3-triazin-3-ylmethyl) O,O-dimethyl phosphorodithioate	Diallate	S(2,3-dichloroallyl)-diisopropylthiocarbamate
Dichlorvos	2,2-dichloro-vinyl dimethyl-phosphate	EPTC	S-ethyldipropylthiocarbamate
Fenchlorphos	dimethyl-2,4,5-trichloro-phenyl phosphorothioate	Noruron	3-(hexahydro-4,7-methanoindan-5-yl)-1,1-dimethyl urea
Mevinphos	2-methoxy-carbonyl-1-methyl-vinyl dimethylphosphate	Sulfallate	2-chloroallyl diethyldithio-carbamate
Monocrotophos	cis-1-methyl-2-methylcarbamoyl-vinyl phosphate	Triallate	S(2,3,3-trichloroallyl)diisopropyl thiocarbamate
Parathion methyl	dimethyl-4-nitrophenyl phosphorothionate		
Trichlorfon	dimethyl-2,2,2,trichloro-1-hydroxyethyl phosphonate		

[a]Morpurgo et al. (1977); Carere et al. (1978b).

only recombinogenic. Although a genetic activity due to the phosphoric moiety is not to be discarded, the results obtained in our experimental conditions suggest that the genetic activity of dichlorvos and trichlorfon is mainly due to the chlorinated part of their molecule.

A similar study was performed with the 5 carbamates; 3 of them, diallate, triallate and sulfallate, contain in their molecule a 2-chloro-allyl group; 2 other carbamates, namely EPTC and noruron, are devoid of the chlorinated group. As reported by Morpurgo et al. (1977) and by Carere et al. (1978), triallate and sulfallate are mutagenic and recombinogenic, whereas completely negative results were obtained with EPTC and noruron. Diallate, negative in *S. typhimurium*, was positive in *S. coelicolor* and *A. nidulans*. It seems reasonable to

TABLE 4

MUTAGENIC AND RECOMBINOGENIC ACTIVITIES OF ORGANOPHOSPHORUS ESTERS AND CARBAMATES[a]

Organophosphorus esters	S. typhimurium (Ames test)	S. coelicolor	A. nidulans gene mut.	A. nidulans mitotic c.o.	A. nidulans mitotic n-d
Azinphos methyl	—	—	—	—	—
Dichlorvos	(+)(TA1535)	+	+	+	+
Fenchlorphos	—	—	—	—	—
Mevinphos	—	—	—	—	—
Monocrotophos	—	—	—	—	—
Parathion methyl	—	—	—	—	—
Trichlorfon	(+)	+	—	+	—
Carbamates					
Diallate	—	(+)	+	+	—
EPTC	—	—	—	—	—
Noruron	—	—	—	—	—
Sulfallate	+(TA1535)	+	+	+	—
Triallate	+(TA1535)	+	+	+	+

[a]Morpurgo et al. (1977); Carere et al. (1978b).

assume that the chlorinated part of their molecules was responsible for the genetic activity of the 3 herbicides. Practically the same conclusions can be drawn from the results obtained by De Lorenzo et al. (1978) and by Sikka et al. (1978) in *S. typhimurium*. De Lorenzo et al. (1978) assayed 20 carbamates with the Ames test. Among them, only 3 (diallate, triallate and sulfallate) containing the 2-chloroallyl group were mutagenic whereas all the others, devoid of the chlorinated group, were negative. A not yet explained point is that, whereas in our experimental conditions diallate, triallate and sulfallate are direct mutagens, in the conditions used by the 2 other authors the 3 herbicides are indirect mutagens.

In our laboratory, Benigni and Dogliotti tested diallate, triallate and sulfallate for their ability to stimulate UDS both with autoradiographic and scintillation-counting techniques; their results showed that the 3 herbicides are positive, with dose—effect curves, only when analyzed with the autoradiographic technique. Diallate (IARC, Vol. 12, 1976) and sulfallate (NCI, Techn. Rep. No. 115, 1978) were carcinogenic. The use of the 3 herbicides is prohibited in Italy.

Another comparative mutational study was carried out on sulfallate and on 4 structurally related halogenated compounds (Fig. 2). 3 saturated short-chain haloalkanes — ethylene dibromide (DBE), ethylene dichloride (DCE), propylene dichloride (DCB) — used in agriculture as insecticidal fumigants, and one unsaturated hydrocarbon — allyl alcohol — containing in its molecule a vinyl group, is used as a herbicide. 3 of these fumigants, DCE (NCI, Techn. Rep. No. 55, 1978), DBE (NCI, Techn. Rep. No. 86, 1978) and sulfallate are carcinogenic in mice and rats.

We tested these compounds for their ability to induce point mutations in *S. typhimurium*, *S. coelicolor* and *A. nidulans* both with spot and plate tests. The spot test, particularly suitable for testing volatile compounds, was the most sensitive one with the Ames test which was applied with strains TA1535,

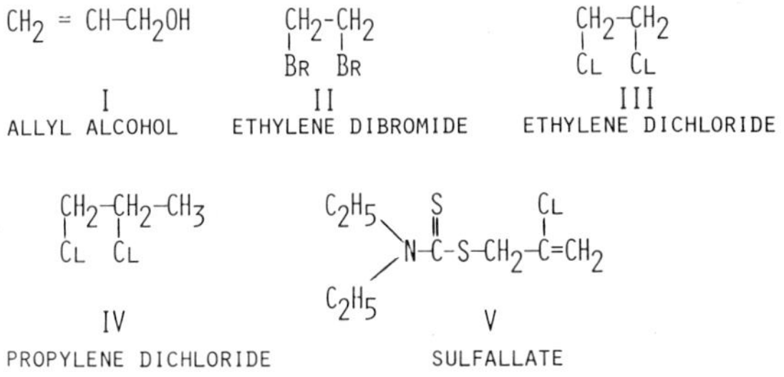

$CH_2=CH-CH_2OH$

I
ALLYL ALCOHOL

CH_2-CH_2
  | |
 Br   Br

II
ETHYLENE DIBROMIDE

CH_2-CH_2
  | |
 Cl   Cl

III
ETHYLENE DICHLORIDE

$CH_2-CH_2-CH_3$
  | |
 Cl   Cl

IV
PROPYLENE DICHLORIDE

C_2H_5
       \\   S   Cl
        N-C-S-CH_2-C=CH_2
       /   ‖
C_2H_5

V
SULFALLATE

Fig. 2. Fumigants analyzed in Salmonella, Streptomyces and Aspergillus for induction of point mutations.

TA1537, TA1538, TA98 and TA100. With this technique it was found (Table 5) that only allyl alcohol was ineffective in reverting all the tester strains; DBE and sulfallate were the most powerful mutagens on the TA1535 and TA100 strains. Sulfallate also reverted the TA98 strain, DCE was weakly positive only on the TA1535 strain in the presence of S9, and DCP exerted a weak effect on the base-substitution type strains. The Ames test was then applied by the quantitative plate-incorporation assay only with TA1535 and TA100 strains (Fig. 3). Allyl alcohol and DCE were completely inactive on both tester strains with and without S9. The mutagenic potency of DBE was partially reduced by the addition of S9 mix, but the genetic activity of sulfallate was not notably affected by the presence of the microsomal fractions.

The *S. coelicolor* and *A. nidulans* forward-mutation systems were applied only without metabolic activation; in both organisms, allyl alcohol and DCE were completely negative. In *A. nidulans*, a weak positive effect was observed

TABLE 5

MUTAGENICITY OF COMPOUNDS TESTED IN *S. typhimurium* SPOT TEST

Compound	μl/plate	his$^+$ revertants/plate[a]									
		TA1535		TA1537		TA1538		TA98		TA100	
		−S9	+S9	−S9	+S9	−S9	+S9	−S9	+S9	−S9	+S9
Allyl alcohol	0	16	17	16	19	9	36	35	43	223	117
	0.05	22	12	14	20	9	33	20	35	160	108
Ethylene di-bromide	0	39	33	17	10	19	21	82	77	185	233
	100	*1250*	*480*	13	6	15	10	114	67	*1800*	*635*
Ethylene di-chloride	0	39	19	17	10	19	21	82	77	188	171
	100	46	58	8	8	12	27	83	84	188	169
Propylene dichloride	0	40	17	16	19	9	36	35	43	223	117
	10	*169*	40	16	27	11	37	45	49	*394*	*235*
Sulfallate	0	13	17	12	12	15	36	27	43	154	117
	10	*700*	*421*	13	20	8	37	*151*	*103*	*2319*	*1431*
Positive[b] controls	−	*1160*	*265*	*465*	*100*	*728*	*580*	*1688*	*835*	*720*	*732*

[a] $\bar{X}$ from 3 plates; S.E. never exceeded 10% of averages.
[b] Positive controls were: ethyl methane sulfonate, 5 μl/plate, for TA1535; 9-aminoacridine, 10 μg/plate, for TA1537; 4-n-o-phenylenediamine, 10 μg/plate, for TA1538 and TA98; methyl methane sulfonate, 1 μl/plate, for TA100; 2-aminoanthracene, 1 μg/plate, for all the strains with S9 Mix. In the table are reported historical values: averages of 12 determinations performed during the whole work.

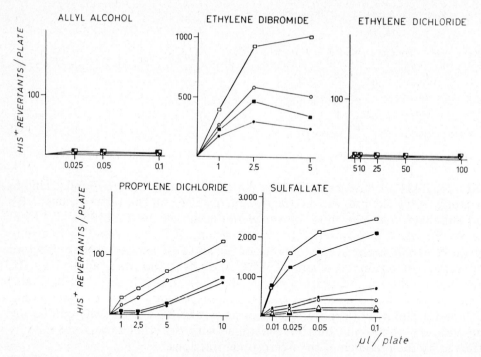

Fig. 3. *S. typhimurium*, plate incorporation assay. ○, TA1535, ●, TA1535 with S9 (50 µl/plate); □, TA100; ■, TA100 with S9 (50 µl/plate). $\bar{x}$ from 3 determinations; SE did not exceed 10% of the averages. Control values have been subtracted; they ranged from 10 to 40 and from 160 to 199 his^+ revertants/plate for TA1535 and TA100, respectively.

with DCP which, on the contrary, was negative in Streptomyces. Sulfallate was positive in both microorganisms. Table 6 summarizes the results obtained in the 3 microorganisms; with these 5 fumigants the *S. coelicolor* system was the less

TABLE 6

COMPARISON OF MUTAGENIC ACTIVITIES OBSERVED IN SALMONELLA, STREPTOMYCES AND ASPERGILLUS

Compound	Dose range (µl/plate)	S. typhimurium		Dose range (µl/plate)	S. coelicolor	Dose range (µl/plate)	A. nidulans
		TA1535	TA100				
Allyl Alcohol	0.025—0.1	—	—	2—100	—	10—40	—
Ethylene Dibromide	1—5	+++	++	2—100	+	10—200	++
Ethylene Dichloride	5—100	—	—	2—100	—	250—500	—
Propylene Dichloride	1—10	+	+	2—100	—	100—400	+
Sulfallate	0.01—0.1	+++	+++	0.1—0.5	+	20—160	+

Results are expressed as increase in mutant numbers or frequencies compared with control values.
—, less than 2-fold increase.
+, 2—5-fold increase.
++, 5—10-fold increase.
+++, more than 10-fold increase.

sensitive; moreover, with compounds that were mutagenic in all 3 microorganisms (i.e. DBE and sulfallate) large quantitative differences were observed, possibly due to different DNA-repair capabilities. The absence of an efficient excision-repair system in Salmonella seems to increase the mutagenic potency of such chemicals.

From the results of this study the following conclusions can be drawn. (1) The 3 haloalkanes studied induce only mutations of the base-substitution type; moreover their genetic potency is in the order DBE, DCP, DCE. (2) Sulfallate is a powerful direct mutagenic agent mainly of the base-substitution type. (3) Allyl alcohol is not mutagenic.

Another investigation was performed on 2 structurally related pesticides, i.e. the bipyridilium herbicides diquat and paraquat. At least 2 main reasons suggested this study to us: (1) the widespread use of both compounds; (2) notwithstanding the considerable amount of data available about the toxicity of paraquat and, to a lower extent, of diquat, the data concerning the genotoxic potential of these compounds were rather scarse and contradictory.

To determine the genotoxic potential of the 2 herbicides we used the following tests. The Ames test (with strains TA1535, TA1537, TA1538, TA98 and TA100) with and without microsomal fractions, resistance to 8-azaguanine in *S. typhimurium* (strains hisG46, TA92, TA1535), repair test in *S. typhimurium* (strains TA1978 and TA1538), gene mutations in *A. nidulans* with 2 different genetic systems (8-azaguanine resistance and methionine suppression), lethal recessives in *A. nidulans* and UDS with the autoradiographic techniques.

A summary of the results (Benigni et al., 1979b) is reported in Table 7. Diquat and paraquat were not mutagenic in the Ames test when assayed by spot, plate and liquid tests, with and without metabolic activation. However, they were positive in *S. typhimurium* in the repair test and in the 8-azaguanine resistance forward-mutation test. Both herbicides were also positive in *A. nidulans* in the induction of gene mutation and lethal recessives. Finally, they were positive in the UDS test. The results of this study suggest the following considerations. Diquat and paraquat are able to induce gene mutations only in forward-mutation systems either in prokaryotic or eukaryotic microorganisms. This activity is enhanced when repair-proficient strains are used. This observation, together with the negative results in the Ames test, suggest that diquat and paraquat may be unable to induce point mutations of the base-substitution and frame-shift types, but able to induce other kinds of damage at gene level (small deletions, cross-links, strandbreaks).

The mechanism by which they are mutagenic is unclear. Ross et al. (1979), by using a system in vitro (mouse lymphoblasts) and the DNA alkaline-elution

TABLE 7

COMPARATIVE MUTATIONAL STUDIES WITH DIQUAT AND PARAQUAT[a]

Compound	*S. typhimurium* his$^+$ revert.	*S. typhimurium* repair test	*S. typhimurium* 8-AG resist.	*A. nidulans*		EUE cells (UDS with autorad.)
				p. mutat.	lethal rec.	
Diquat	−	+	+	+	+	+
Paraquat	−	+	+	+	+	+

[a]Benigni et al. (1979b).

technique, found that paraquat is able to cause damage to DNA in the form of single-strand breaks; as a possible mechanism these authors suggest an indirect action on DNA by free radicals or other toxic agents such as H_2O_2 or lipid peroxides.

Our group was then involved in a mutagenicity study with the herbicide trifluralin and 3 trifluorotoluene derivatives detected as environmental contaminants of the water in some wells of a large area in north-eastern Italy, near Vicenza. Gas-chromatographic and mass-spectrometric analyses (Belsito et al., 1979) indicated the following 3 main pollutants as responsible for the contaminations: 4-chlorotrifluorotoluene (CTT), 4-chloro-3-nitrotrifluorotoluene (NCTT) and 4-chloro-3,5-dinitrotrifluorotoluene (DNCTT).

Responsible for this pollution was a chemical plant that for many years had been producing several fluorotoluene derivatives among which, on the pilot scale, was DNCTT, an intermediate in the synthesis of dinitroaniline herbicides such as trifluralin (trifluoro-2,6-dinitro-*N*,*N*-dipropyl-*p*-toluidine). The 3 intermediates, as well as trifluralin, were assayed with 3 tests to detect gene mutations (the Ames test), mitotic segregation (with selective and unselective methods in *A. nidulans*) and UDS with autoradiography. Table 8 summarizes the results: in our experimental conditions the 4 compounds failed to exert any mutagenic activity in the Ames test with and without microsomal activation; the 3 trifluorotoluenes were negative in *A. nidulans*, when tested for somatic segregation both with spot and plate tests. Only trifluralin (Table 9) induced mitotic crossing-over in repeated experiments, however without a dose—effect curve, probably due to the saturation of trifluralin solutions. Trifluralin was tested for non-disjunction induction with a liquid test applied to germinating conidia. The preliminary results obtained in one experiment (Table 10), which seem to suggest a weak effect on non-disjunction together with a weak effect on crossing-over (Table 9), require confirmation with further experiments of higher dimensions. In conclusion, of the 3 pollutants, only CTT and NCTT showed a genotoxic activity, and then only in one test, the UDS. Trifluralin, which was reported by the NCI as carcinogenic in female mice (Techn. Rep. No. 34, 1978), in our studies induced no point mutations or UDS;

TABLE 8

COMPARATIVE MUTATIONAL STUDIES WITH TRIFLURALIN AND TRIFLUOROTOLUENE DERIVATIVES

Compound	S. typhimurium (Ames test)	A. nidulans		EUE cells (UDS with autorad. technique)
		mitotic c.o.	mitotic n-d	
4-Chloro-trifluoro-toluene (CTT)	—	—	—	+
4-Chloro-3-nitro-trifluorotoluene (NCTT)	—	—	—	+
4-Chloro-3,5-dinitro-trifluorotoluene (DNCTT)	—	—	—	—
Trifluralin	—	+	(—)	—

TABLE 9

INDUCTION OF MITOTIC RECOMBINATION (FPA RESISTANCE) IN Aspergillus nidulans[a]

Compound tested	mg/plate	FPA resistants/plate $\bar{x} \pm SE$
Control	0	2.62 0.73
CTT	0.25	3.00 0.40
	1.00	2.00 0.91
	2.50	2.00 0.40
NCTT	0.25	1.75 0.47
	0.50	2.25 0.75
	1.00	1.50 0.64
DNCTT	0.01	3.75 0.47
	0.02	2.75 0.47
	0.05	2.00 0.91
Trifluralin	0.10	14.25 5.70
	0.50	12.00 1.15
	1.00	10.00 1.60
	5.00	17.25 3.50

[a]The technique applied was the spot test.

TABLE 10

INDUCTION OF MITOTIC NON-DISJUNCTION IN ASPERGILLUS BY TRIFLURALIN[a]

Concentration (% w/v)	Total number of colonies tested	n-d colonies	
		Number	freq/10^3
0	1484	3	2.0
0.05	544	3	5.5
0.10	564	6	10.6
0.50	900	2	2.2
1.00	890	3	3.4

[a]The technique used was a liquid test applied to germinating conidia.

however, in contrast with previously reported results observed in S. cerevisiae (Simmon et al., 1977), it was able to induce a significant increase of mitotic crossing-over.

The last point of this paper concerns the results obtained with a study of the effects of plant metabolism. The measurement of the potential mutagenicity of pesticides is complicated because pesticides may be metabolized by plants. It is therefore useful to devise methods to verify whether the plant metabolism transforms an initially harmless pesticide into a mutagenic product and vice versa. To this aim a simple and rapid method, based on Nicotiana alata cell cultures in vitro, was set up in our laboratory. The Nicotiana cells may easily be cultivated in suspension in large test tubes on revolving drums in a thermostatic room at 24°C. The pesticides may be added to the cell cultures at various concentrations; the homogenate may then be used directly in petri dishes with the mutagenicity tests. The cells of the test organism may be incubated with the homogenized N. alata cells and then analyzed for the appearance of mutations. However, to be able to use such a method in vitro one must prove that the vegetable cells in vitro simulate the metabolism occurring in the field. To this aim we have so far tested the effects of the Nicotiana cell cultures in vitro on 5 pesticides — atrazine, dichlorvos, tetrachlorvinphos, kelevan and maleic hydrazide — belonging to different chemical classes and for which there is knowledge of their metabolism.

The results of this study were published by Benigni et al. (1979a). The results shown in Table 11 were obtained, with gas-chromatographic and chromatographic techniques, on the metabolic effects of Nicotiana cells on the 5 pesticides tested. In the case of atrazine the metabolic rate was increased with time by 74% after 30 days. Dichlorvos was metabolized to a great extent (78% in 21 days); tetrachlorvinphos was completely metabolized after 12 days. On the contrary, maleic hydrazide, in accord with what is known in the field, was not metabolized. Kelevan was transformed into kepone slightly during the first 7 days but more after 10—15 days. In conclusion, at least in the case of the five pesticides studied, the use of *N. alata* cell cultures in vitro simulates quite well the plant metabolism occurring in the field. The time necessary to perform one treatment is no longer than 1 month at the most, and the accuracy and reproducibility are satisfactory.

As to the mutagenicity studies performed with the 5 pesticides with and without Nicotiana metabolism, only atrazine turned out to be positive in the induction of gene mutations (8-azaguanine resistance) and mitotic crossing-over analyzed with a liquid test procedure. The results obtained with atrazine are reported in Table 12.

TABLE 11

METABOLISM OF *N. alata* CELL CULTURE WITH SOME PESTICIDES[a]

Compound	Treatment time (d)	Amount/tube (mg)	Recovery from medium (mg)	Recovery from medium + cells	Metabol. (%)
Atrazine	8	1	1.00	0.80	20
	15	1	0.87	0.45	48
	22	1	0.83	0.24	71
	30	1	0.83	0.22	74
Dichlorvos	3	4	4.0	4.0	0
	7	4	3.6	2.0	45
	14	4	3.0	1.0	67
	21	4	2.3	0.5	78
Tetrachlorvinphos (Gardona)	12	2	1.6	0	100
Maleic hydrazide	12	2	1.5	1.41	6
			Recovery of kelevan + kepone (mg)	Recovery of kepone (mg)	
Kelevan	7	0.125	0.10	0.02	
	7	0.50	0.40	0.10	
	7	1.00	1.00	0.00	
	7	2.00	1.80	0.30	
	7	1.00	0.90	0.15	
	10	1.00	0.90	0.30	
	15	1.00	0.80	0.50	

[a] Benigni et al. (1979a).

TABLE 12

MUTAGENIC AND RECOMBINOGENIC ACTIVITY OF ATRAZINE IN A. nidulans AFTER PLANT METABOLISM[a]

	Amount/pl. (mg)	Colonies/pl. $\bar{x} \pm SE$	Surv. (%)	Mutation or segregation frequency
		8-Azaguanine resistance		
Atrazine	0	1.83 ± 0.79	100	0.37 × 10^6
	1	2.00 ± 0.44	100	0.40 × 10^6
	2	1.33 ± 0.21	100	0.27 × 10^6
	5	1.50 ± 0.22	62	0.30 × 10^6
Atrazine plus	0	5.0 ± 0.36	100	1.0 × 10^6
Nicotiana cell	2	4.7 ± 0.44	100	0.9 × 10^6
medium	4	2.5 ± 0.25	92	0.5 × 10^6
Atrazine after	control	1.0 ± 0.36	100	0.4 × 10^6
plant metabol.	treated	3.9 ± 0.96	33	2.3 × 10^6
		Somatic segregation		
Atrazine plus	0	6.0 ± 0.26	100	1.2 × 10^4
Nicotiana cell	2	6.8 ± 0.46	100	1.4 × 10^4
medium	4	5.8 ± 0.41	89	1.3 × 10^4
Atrazine after	control	3.8 ± 0.96	100	0.76 × 10^4
plant metabol.	treated	6.5 ± 1.19	34	3.80 × 10^4

[a] Benigni et al. (1979a).

Conclusions

More than half (21 out of 37) of the pesticides investigated were positive in at least one genetic system. 4 of them (benomyl, CTT, NCTT, picloram) were positive only in one genetic system. The very high proportion of positive results can be explained because many of the pesticides assayed were selected on the basis of suspicion (several of them were known to be mutagenic or carcinogenic) and also because pesticides are intended to be biologically reactive compounds. We should mention that, in the long-term bioassay program conducted by the NCI, a rather high proportion (38%) of positive results was obtained (U. Saffiotti, personal communication).

On the basis of the IARC and NCI evaluations, for 8 of the 37 pesticides tested in our programme, there is evidence of carcinogenic activity in at least one species of rodent (Table 13). Among the 8 carcinogenic pesticides, 7 were mutagenic in at least one genetic system.

In conclusion, considering the over-all results obtained in our comparative mutagenicity study we think we may write as follows. (1) A single test system, even if highly sensitive, as for instance the Ames test, is not sufficient to fulfil the difficult task of detecting potential mutagens and/or carcinogens. (2) It is important to use both forward- and back-mutation systems for the induction of gene mutations. (3) Particularly interesting is the analysis of genetic events such as mitotic crossing-over and non-disjunction in A. nidulans, although the extrapolation of non-disjunction to higher organisms is still to be elucidated. (4) Plant metabolism may sometimes be responsible for the mutagenic activity in vitro, as in the case of atrazine. (5) It is important although rather laborious

TABLE 13
GENETIC ACTIVITY OF PESTICIDES IDENTIFIED AS CARCINOGENIC IN AT LEAST ONE SPECIES (DATA FROM NCI AND IARC)

Pesticide	Carcinogenesis				Mutagenesis			
	NCI Techn. Rep.	IARC vol.	Y	Species	S. typhimurium	S. coelicolor	A. nidulans	UDS (autorad.)
Aminotriazole (amitrole)	7		74	MR	−	+	(+) g.mutat. mit.c.o.	+(unpubl.)
Captan	15		78	M	+	−	+ g.mutat. mit.c.o.	NT
Diallate[a]		12	76	M	−[a]	+	+ g.mutat. mit.c.o.	+(unpubl.)
Ethylene dibromide	86		78	MR	+	+	+ g.mutat.	NT
Ethylene dichloride	55		78	MR	(+)	−	− g.mutat.	NT
Sulfallate	115		78	MR	+	+	+ g.mutat. mit.c.o.	+(unpubl.)
Trifluralin	34		78	M	−	−	+ mit.c.o.	−
Tetrachlor-vinphos	33		78	M	NT	NT	− g.mutat. mitot.c.o.	(+)(unpubl.)

[a]Diallate was found positive in the Ames test by De Lorenzo et al. (1978) and by Sikka et al. (1978). Y, year; M, mice; R, rats; g.mutat, gene mutations; mit.c.o., mitotic crossing-over.

to use aspecific tests such as UDS applied by autoradiography. (6) The use of a battery of short-term tests is indispensable for the evaluation of the genetic potential of chemicals and useful for the prediction of their carcinogenicity. (7) Our results, together with those of long-term carcinogenicity programs, show how important pesticides are from the toxicological point of view.

Acknowledgements

The experimental data presented were obtained within the framework of a 5-year mutagenicity program partially supported by the Commission of the European Communities (E.E.C.) (projects Nos. 069-74-1 and 177-77-1 ENV I). Responsible for the chemical part of the programme was Prof. I. Camoni.

The results were obtained thanks to the excellent collaboration, at different times, of the following: Drs. D. Bellincampi, R. Benigni, M. Bignami, G. Cardamone, P. Comba, R. Crebelli, E. Dogliotti, E. Falcone, G. Gualandi, R. Iachetta, V.A. Ortali, P. Principe, A. Velcich and Messrs. A. Calcagnile, O. Cervelli, G. Conti, L. Conti and A. Novelletto. We are grateful to our collaborators for useful discussions and permission to quote some of their unpublished results.

References

Ames, B.N., F.D. Lee and W. Durston (1973) An improved bacterial test system for the detection and classification of mutagens and carcinogens, Proc. Natl. Acad. Sci. (U.S.A.), 70, 782—786.

Ames, B.N., J. McCann and E. Yamasaki (1975) Methods for detecting carcinogens and mutagens with Salmonella/mammalian-microsome mutagenicity test, Mutation Res., 31, 347—364.

Belsito, F., L. Boniforti, R. Dommarco and G. Laguzzi (1979) Mass spectra and fragmentation patterns of 4-chloro-trifluoro-toluene, 4-chloro-3-nitro-trifluoro-toluene and 4-chloro-3,5-dinitro-trifluorotoluene, Ann. Chem., 69, 259.

Benigni, R., M. Bignami, I. Camoni, A. Carere, G. Conti, R. Iachetta, G. Morpurgo and V.A. Ortali (1979a) A new in vitro method for testing plant metabolism in mutagenicity studies, J. Toxicol. Environ. Health, 5, 809—819.

Benigni, R., M. Bignami, A. Carere, G. Conti, L. Conti, R. Crebelli, E. Dogliotti, G. Gualandi, A. Novelletto and V.A. Ortali (1979b) Mutational studies with diquat and paraquat in vitro, Mutation Res., 68, 183—193.

Bignami, M., and R. Crebelli (1979) A simplified method for the induction of 8-azaguanine resistance in Salmonella typhimurium, Toxicol. Lett., 3, 169—175.

Bignami, M., G. Morpurgo, R. Pagliani, A. Carere, G. Conti and G. Di Giuseppe (1974) Non-disjunction and crossing-over induced by pharmaceutical drugs in Aspergillus nidulans, Mutation Res., 26, 159—170.

Bignami, M., F. Aulicino, A. Velcich, A. Carere and G. Morpurgo (1977) Mutagenic and recombinogenic action of pesticides in A. nidulans, Mutation Res., 46, 395—402.

Bignami, M., G. Conti, L. Conti, R. Crebelli, F. Misuraca, A.M. Puglia, R. Randazzo, G. Sciandrello and A. Carere (1980) Mutagenicity of halogenated aliphatic hydrocarbons in Salmonella, Streptomyces and Aspergillus, Chem.-Biol. Interact., 30, 9—23.

Carere, A., V.A. Ortali, G. Cardamone, A. Torracca and R. Raschetti (1978a) Microbiological mutagenicity studies of pesticides in vitro, Mutation Res., 57, 277—286.

Carere, A., V.A. Ortali, G. Cardamone and G. Morpurgo (1978b) Mutagenicity of Dichlorvos and other structurally related pesticides in Salmonella and Streptomyces, Chem.-Biol. Interact., 22, 297—308.

De Lorenzo, F., N. Staiano, L. Silengo and R. Cortese (1978) Mutagenicity of Diallate, Sulfallate and Triallate and relationship between structure and mutagenic effects of carbamates used widely in agriculture, Cancer Res., 38, 13—15.

Fratello, B., G. Morpurgo and G. Sermonti (1960) Induced somatic segregation in Aspergillus nidulans, Genetics, 45, 785.

Hastie, A.C. (1971) Benlate induced instability of Aspergillus diploids, Nature (London), 226, 771.

Kappas, A., S.G. Georgopoulos and A.C. Hastie (1974) On the genetic activity of benzimidazole thiphanate fungicides on diploid Aspergillus nidulans, Mutation Res., 26, 17—27.

Lilly, L.J. (1965) An investigation on the suitability of suppressors of methA1 in Aspergillus nidulans for the study of induced and spontaneous mutation, Mutation Res., 2, 192—195.

Morpurgo, G., F. Aulicino, M. Bignami, L. Conti and A. Velcich (1977) Relationship between structure and mutagenicity of Dichlorvos and other pesticides, Accad. Naz. Lincei, Ser. VIII, Vol. 63, No. 5, 693—701.

Morpurgo, G., S. Puppo, G. Gualandi and L. Conti (1978) A quick method for testing recessive lethal damage with a diploid strain of *Aspergillus nidulans*, Mutation Res., 54, 131—137.

Morpurgo, D., D. Bellincampi, G. Gualandi, L. Baldinelli and O. Serlupi-Crescenzi (1979) Analysis of mitotic non-disjunction with *Aspergillus nidulans*, Environ. Health Perspect., 31, 81—95.

Ross, W.E., E.R. Block and R.Y. Chang (1979) Paraquat induced DNA damage in mammalian cells, Biochem. Biophys. Res. Commun., 91, 1302—1308.

San, R.H.C., and H.F. Stich (1975) DNA-repair synthesis of cultured human cells as a rapid bioassay for chemical carcinogens, Int. J. Cancer, 16, 284—291.

Shopek, T.R., H.L. Liber, J.J. Krolewski and W.G. Thilly (1978) Quantitative forward mutation in *Salmonella typhimurium* using 8-azaguanine resistance as a genetic marker, Proc. Natl. Acad. Sci. (U.S.A.), 75, 410.

Sikka, H.C., and P. Florczyk (1978) Mutagenic activity of thiocarbamate herbicides in *Salmonella typhimurium*, J. Agric. Food Chem., 26 No. 1, 146—148.

Simmon, V.F., A.D. Mitchel and T.A. Jorgenson (1977) EPA 600/1-77-023.

CASE-CONTROL STUDIES: SOFT-TISSUE SARCOMAS AND MALIGNANT LYMPHOMAS AND EXPOSURE TO PHENOXY ACIDS OR CHLOROPHENOLS

L. HARDELL

Department of Oncology, University Hospital, S-901 85 Umeå (Sweden)

Chemical pesticides have been used in agriculture and forestry in Sweden since the end of the 1940's. Whereas the consumption of insecticides has declined during the last 25 years and that of fungicides remained more or less constant, the use of herbicides has increased markedly.

Of the 4394 ton of active pesticides sold to agriculture in 1977, 3776 ton (86%) consisted of herbicides. Of the herbicides, the phenoxy acids amounted to 2690 ton (71%). Besides being present in most of the common agricultural herbicides, phenoxy acids are used in horticulture and forestry, as well as to control unwanted hardwoods, as in the killing of individual trees, i.e. basal bark spraying.

During the last few years, exposure to phenoxy acids has attracted an increasing interest as a possible cause of malignant disorders. Case reports on soft-tissue sarcomas (Hardell, 1977) and malignant lymphomas (Hardell, 1979), as possibly related to exposure to phenoxy acids have strengthened the suspicions about a carcinogenic effect of these preparations.

Stimulated by these observations two case-control studies on soft-tissue sarcomas (Hardell and Sandström, 1979; Eriksson et al., 1980) and malignant lymphomas (Hardell et al., 1980) were performed. An analysis of exposure to chlorophenols was included in these investigations, as there are related processes in the production of phenoxy acids and chlorophenols. Furthermore, there may be similar impurities such as chlorinated dibenzodioxins and dibenzofurans in phenoxy acids and chlorophenols. Chlorophenols are mainly used to protect against blue stain and as impregnates in the saw mill industry, but are also used as wood preservatives in tanneries, and as proofing in clothes, etc.

An about 5-fold increase in the risk for soft-tissue sarcomas and exposure to phenoxy acids or chlorophenols was found in the soft-tissue sarcoma studies, which were performed in two areas in Sweden (Table 1). An association was also shown between malignant lymphomas and exposure to phenoxy acids (Table 2), chlorophenols (Table 3), but also to organic solvents (Table 4).

Each of the 3 case-control studies were designed to minimize the likelihood

TABLE 1

RISK RATIOS FOR SOFT TISSUE SARCOMA IN TWO CASE-CONTROL STUDIES

Exposure category	Study location	
	Northern Sweden[1]	Southern Sweden[2]
Phenoxy acids and/or chlorophenols	5.7(6.2)[a]	4.7(5.1)[a]
Phenoxy acids only	5.3	6.8
Chlorophenols only	6.6	3.3[b]

[a] Estimates in parentheses were calculated from matched material. All others were derived from unmatched material.
[b] $p < 0.01$. For all other relative risks, $p < 0.001$.
Sources: (1) Hardell and Sandström (1979).
(2) Eriksson et al. (1980).

TABLE 2

EXPOSURE TO PHENOXY ACIDS

Cases/Referents	Exposed			Unexposed
	≥ 90 days	< 90 days	Total	
Cases	10	31	41	108
Referents	4	20	24	303
Risk ratio	7.0	4.3	4.8	(1.0)

TABLE 3

EXPOSURE TO CHLOROPHENOLS

Cases/Referents	Unexposed	Exposed		
		Low grade[a]	High grade	Total
Cases	94	14	25	39
Referents	284	19	9	28
Risk ratio	(1.0)	2.2	8.4	4.2

[a] A continuous exposure for no more than one week or repeated brief exposure totaling at most one month was classified as low-grade.

TABLE 4

EXPOSURE TO ORGANIC SOLVENTS

Cases/Referents	Unexposed	Exposed		
		Low-grade[a]	High-grade	Phenoxy acids and/or chlorophenols
Cases	60	10	40	23
Referents	222	31	47	10
Risk ratio	(1.0)	1.2	3.1	8.5

[a] A continuous exposure for no more than one week or repeated brief exposure totaling at most one month was classified as low-grade.

of spurious or misleading results. In the first soft-tissue sarcoma study (Hardell and Sandström, 1979) and in the lymphoma study (Hardell et al., 1980) the cases consisted of individuals who had been admitted to the Department of Oncology in Umeå. Cases in the second soft-tissue sarcoma study (Eriksson et al., 1980) were obtained from the Cancer Registry of the Swedish Social Welfare Board and were patients who resided in a particular region of Southern Sweden at the time of diagnosis.

Controls were selected by a method designed to minimize the influence of confounding factors. In each of the studies controls were matched individually to each case for sex, age and place of residence. In addition, decreased controls were selected for deceased patients and matched for year of death.

In each of the studies, exposure information was obtained by written questionnaires and was supplemented by telephone interviews. By such a method the individuals' attitudes and previous experiences could introduce bias into the assessment of exposure. One possibility could be a tendency, conscious or not, for those suffering from a disease to overstate past exposure as compared to healthy controls. Each study was designed to minimize such recall bias. Each subject or next-of-kin of a deceased subject received by mail a questionnaire which contained a variety of questions concerning previous and present occupation, conditions in the occupational environment, smoking habits, etc., avoiding any special attention to phenoxy acid or chlorophenol exposure. Supplementary interviews were carried out by individuals who did not know whether the subjects were cases or controls. In order to avoid any bias which might result from differences in recall between live subjects and deceased subjects' next-of-kin deceased controls were selected for deceased patients.

To analyse possible observational bias the subjects were divided between those employed in occupations where exposure to phenoxy acids could be expected (agriculture and forestry) and those employed in other occupations. Separate relative risks could then be calculated for exposed and unexposed workers in agriculture and forestry as compared to unexposed persons in other occupations (Table 5). If cases were inclined to substantially exaggerate their exposure, this bias would increase the number of exposed cases in occupations where exposure is commonly creating an artificial deficit in unexposed cases in these occupations. Thus, the presence of substantial recall bias should result in a relative risk considerably below 1.0 when unexposed individuals in agriculture and forestry are compared to unexposed individuals in other occupations. However, the obtained relative risks (1.1 resp. 0.9) indicated that bias, if present, did not substantially influence the results.

TABLE 5

EXPOSURE TO PHENOXY ACIDS DIVIDED BY OCCUPATION
Relative risks calculated within agriculture/forestry.

	Agriculture/forestry		Other occupations
	Exposed	Unexposed	Unexposed
Soft-tissue sarcoma study (southern Sweden)	6.4	1.1	(1.0)
Malignant lymphoma study	4.1	0.9	(1.0)

In order to further evaluate any possible observational bias in the soft-tissue sarcoma and lymphoma studies another similar investigation was made. This time the cases constituted of patients with colon cancer. This type of cancer has not previously been suspected to be associated with exposure to phenoxy acids or chlorophenols. Discussion of phenoxy acids and their presumptive risks was intense in Sweden during the period when this new study was conducted. Accordingly any observational bias present in the first 3 studies should also be present in this study. Analysis indicates however that there is no association between colon cancer and exposure to phenoxy acids. Consequently, any substantial observational bias which could give the shown relative risks in the previous studies does probably not exist.

In summary, the investigations indicate an association between exposure to phenoxy acids and chlorophenols and soft-tissue sarcomas and malignant lymphomas.

References

Eriksson, M., L. Hardell, N.O. Berg, T. Möller and O. Axelson, (1980) Soft-tissue sarcomas and exposure to chemical substances, A case-referent study, Br. J. Ind. Med., in press.

Hardell, L. (1977) Soft-tissue sarcomas and exposure to phenoxy acetic acids — a clinical observation, Läkartidningen, 74, 2753—2754.

Hardell, L. (1979) Malignant lymphoma of histiocytic type and exposure to phenoxyacetic acids or chlorophenols, Lancet, 1, 55—56.

Hardell, L., and A. Sandström (1979) Case-control study: Soft-tissue sarcomas and exposure to phenoxyacetic acids or chlorophenols, Br. J. Cancer, 39, 711—717.

Hardell, L., M. Eriksson, P. Lenner and E. Lundgren (1980) Malignant lymphoma and exposure to chemical substances, especially organic solvents, chlorophenols and phenoxy acids, A case-control study, Br. J. Cancer, in press.

MUTAGENIC AND CARCINOGENIC POLLUTANTS

CIGARETTE SMOKE INDUCED DNA DAMAGE IN MAN

H.J. EVANS

MRC Clinical and Population Cytogenetics Unit, Western General Hospital, Crewe Road, Edinburgh (Great Britain)

Experimental studies on laboratory animals (Dontenwill et al., 1973) and epidemiological studies on our own species (Royal College of Physicians, 1977) leave no doubt that cigarette smoking is the cause of almost all cases of anaplastic and squamous cell bronchial carcinoma in man. Cigarette smoking is a major cause of premature death in modern society, since not only is it associated with the development of lung cancers, but also with cancers of the oral cavity, of the larynx, oesophagus, pancreas and bladder; with coronary heart disease and stroke; with bronchitis and emphysema; and even peptic ulcer (WHO, 1975). The list is long so that the conclusion that heavy cigarette smokers in modern society might live, on average, for some 10—15 years less than their non-smoking counterparts, is not surprising. It is also well-known that cigarette smoking during pregnancy has marked detrimental effects upon the growth and development of the unborn child and it has been recently demonstrated that adolescents who were exposed to tobacco smoke products whilst fetuses in the wombs of their mothers, show reduced IQs relative to controls born of non-smoking mothers. Moreover, it has been possible to detect and demonstrate small airway dysfunction in non-smokers who breathe the smoke produced by the cigarettes smoked by someone else (White and Froeb, 1980). In other words, it appears that passive smoking may have possible deleterious consequences.

How do all these diverse consequences of cigarette smoking relate to cigarette smoke induced DNA damage in man? Some of these effects are almost certainly a consequence of smoke, or smoke product, induced delays in cell proliferation and to cell killing, and these may themselves, at least in part, be a consequence of damage to genetic materials. Of more direct relevance, however, is the fact that cigarette smoking clearly causes cancer and there is considerable evidence to indicate that the transformation to malignancy may be associated with, and indeed may depend upon, damage to DNA.

The relation between cigarette smoking and human lung cancer was originally based on epidemiological evidence produced by Doll and Hill and by Wynder and Graham some 30 years ago. Since then it has been shown that the cancer risk is a function of the number and type of cigarettes smoked (Doll and Peto, 1978) and that cigarette smoke may act synergistically with other cancer-inducing agents. For example, smokers working in metalliferous mines where there are high concentrations of radon or plutonium, and so high levels of radiation exposure (Doll, 1977), or smokers working in the presence of asbestos fibres (Hammond and Selikoff, 1973), are at very much greater risk than non-smokers in these hazardous environments, and similar synergisms have been noted for the development of oral cancers in smokers consuming reasonably large quantities of alcohol (Obe, these proceedings).

Thus, with regard to the induction of human cancers by cigarette smoke, there are 3 general points that could be relevant to smoke-induced DNA damage. These are: (i) That the level of damage, here expressed as the cancer risk, is a function of level of exposure. In other words type of cigarette, number smoked, degree of inhalation, etc. (ii) That cigarette smoke may act in concert with other environmental agents to give rise to the basic biological effects that are of interest to us. (iii) Since a whole range of different body tissues and systems may be affected with cigarette smoke related disease, then this must mean that smoke inhalation into our respiratory system is an efficient route leading to the systemic distribution of soluble and insoluble particulate matter that is initially drawn into the lungs.

The relationship between cancer and DNA damage is a well established one. Cell biologists and geneticists interested in the effects of carcinogens on cells have known for many years that a number of substances that were carcinogens in the mouse were important chromosome-damaging agents in mammalian cells and potent mutagens in cells of plants and in Drosophila. This association between carcinogenicity and mutagenicity has been particularly emphasized more recently by the work of Bruce Ames and his colleagues, and it is not surprising to note that Keir, Yamasaki and Ames (1974) were the first to show that cigarette smoke condensates were, after metabolic activation, mutagenic to the bacterium Salmonella. This first demonstration of the mutagenicity to bacteria of cigarette tobacco products was confirmed by Hutton and Hackney (1975) and later by a number of Japanese workers. It was further shown in these studies that most of the mutagenic activity was in the basic fraction of the smoke condensate, with no detectable activity in the neutral fractions containing the polycyclic aromatic hydrocarbons, but we shall return to this later.

What I want to emphasise at this point is that the association between carcinogenicity and mutagenicity is so close that we might expect that the possible induction of mutations in man as a consequence of exposure to cigarette smoke, might parallel the known induction of cancer in our species as a consequence of cigarette smoking. Cigarette smoke has, therefore, long been considered to be a potential human mutagen and since in many Western populations some 20—50% of the population may be smokers, and since, as we have already pointed out, the active products of tobacco smoke appear to be disseminated throughout the body, then large numbers of people, large amounts of DNA and a substantial fraction of the population gene pool may well be exposed to the mutagenic action of cigarette smoke. In the last few years there have been various pieces of evidence, most of it indirect, which point strongly to the fact that cigarette smoking may well be a mutagenic passtime and I should like to summarise this evidence, as well as some recent relevant work from my own laboratory, under 3 headings.

1. Studies on spermatozoa in cigarette smokers and non-smokers

It has been known for around 20 years that the exposure of spermatogonial cells of the mouse to quite low doses of X-rays, of around 30 rad or so, results in an eventual increase in the proportion of morphologically abnormal sperm present in the ejaculate when the irradiated gonial cells, or rather their descendants, mature into spermatozoa some 2 weeks or so after irradiation (Oakberg and DiMinno, 1960). The frequency of abnormally shaped sperm in inbred mouse strains, and F1 hybrids, is quite low, of the order of 1 or 2%, but this frequency increases markedly after exposure to X-rays and the increase is dose-dependent. More recently, Wyrobek and Bruce (1975) have shown that the incidence of abnormal shaped sperm in the mouse also increases following exposure of the animals to a wide variety of chemical mutagens. Although it has been shown that these sperm abnormalities are not simply a reflection of the presence of induced chromosome aberrations, it is generally believed that at least some of these abnormalities may reflect genetic changes in genes responsible for spermatogenesis, and assaying for sperm abnormalities has now become a reputable technique for screening for substances with possible carcinogenic/mutagenic activities. In this context the work of Viczian (1969) who studied spermatozoa in seminal fluid samples from 120 men who had smoked cigarettes for more than one year and in 50 matched non-smoking controls, would seem to be of particular relevance and some of his findings are summarised in Fig. 1.

The data on abnormal sperm morphology in Fig. 1 show that non-smokers appear to have a frequency of abnormal sperm of less than 20%, whereas those who smoked around 10 cigarettes or less a day had a slightly increased frequency of abnormality, and this went up stepwise with increasing cigarette consumption. As a whole, the smokers showed a higher frequency of abnormality than non-smokers and there is a very clear correlation with the number of cigarettes smoked per day, or rather perhaps I should say here with the

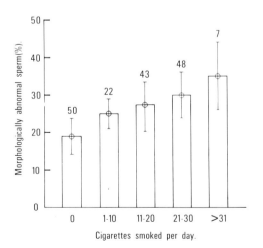

Fig. 1. Proportions of sperm that are morphologically abnormal in samples from 50 non-smokers and 120 smokers. (After Viczian, 1969)

number of cigarettes that the participants said they smoked per day. We should emphasize here that one of the problems in work on relating effects to amount of cigarettes consumed is the question of dosimetry, since people inhale different amounts of smoke, smoke cigarettes of different tar contents, and often claim they smoke fewer cigarettes than they actually do; and this is a problem we shall also return to later.

The Viczian data have been quoted a number of times by people interested in looking to see whether cigarette smoking has an effect on the germ cells of man (ICPEMC, 1979), but they are somewhat unusual in two respects. First, the exceedingly good correlation between degree of abnormality and apparent numbers of cigarettes smoked and, second, by the unusually low incidence of abnormal shaped sperm in non-smokers. In this latter context, I should mention that we have over the past 20 years in our Unit been routinely analysing sperm samples from a range of males attending a clinic for the sub-fertile. Since sub-fertility is also as much, or often more, a fault on the female side, then a large number of males whom we examine are perfectly normal and fertile individuals and their sperm samples are normal. Nevertheless, as in every other laboratory that I am familiar with which analyses human sperm, we find that normal human virile fertile males who do not smoke rarely show more than 60—70% of normal well-shaped sperm. In most laboratories then the frequency of morphologically abnormal sperm in non-smokers is higher than that found by Viczian in his smokers.

This high incidence of sperm abnormalities in man has been a considerable puzzle and some years ago it was suggested that the reason for this high frequency of abnormal spermatozoa in men and low frequency in wild animals such as the mouse etc., was simply because men wore "Y-fronts" so that their testicles were enclosed in an environment which provided too high a temperature (Bedford et al., 1973). An opportunity to test this "hyperthermia hypothesis" came up when we had the chance of looking at sperm samples from some gorillas in the zoo. In this study, and somewhat to our surprise, we found that the gorilla also has a high incidence of abnormal sperm, around 30—40% just as in man although the chimpanzee and orang utang did not (Seuanez et al., 1977). Hyperthermia and cigarette smoking are therefore not the cause of the high incidence of sperm abnormalities in man or gorilla, but the Viczian data and the fact that exposure to chemical mutagens increases the incidence of abnormal spermatozoa in mouse, indicated that a further study of cigarette smoking on human spermatozoa was warranted.

In collaboration with my colleagues Mr. Hargreave and Mrs. Fletcher we therefore had a look at sperm counts, motility, morphology, etc. from a series of individuals who were attending our sub-fertility clinic and who showed a range of sperm count and amongst many of whom were men of known fertility. All slides were scored without knowledge of smoking habits or otherwise of any of the individuals, and any data from people with varicoceles was excluded. After the scoring of a few hundred individuals we then selected 50 smokers and 50 matched non-smokers without knowledge of their sperm profiles except for sperm count. The matching was then made on the basis of sperm count and age and details of our findings will be published elsewhere (Evans et al., 1981). In brief, we once again find that fully fertile vigorous males show around 30—40% or so of abnormally shaped sperm, but if we categorise individuals into groups based on the proportions of abnormal sperm, then there is a significant excess of smokers in the classes with the highest numbers of abnormal sperm and the difference in the distribution between smokers and non-smokers is highly significant. The mean frequency of normal sperm is 60% in non-smokers and 53% in smokers — a difference that is significant at the 1% level. We have also analysed our data by pairing smokers and non-smokers with equal sperm counts: 41 comparisons again give a significant excess of abnormals in the smokers at the 1% level of significance. Our data therefore confirm Viczian's conclusion that the frequency of abnormally shaped sperm is higher in cigarette smokers than in non-smokers. However, when we partition our data on the basis of the number of cigarettes smoked per day then, unlike Viczian, we can find no correlation between dose and effect. But as I mentioned before, measurement of dose in terms of inhaled cigarette smoke is fraught with problems.

All we can conclude from this is that there clearly is an effect of cigarette smoke in producing an increased frequency of sperm abnormalities in man. If we can extrapolate from what we know about some of the causes of sperm abnormalities in the mouse, then it is possible that this increase may reflect an increase in genetic damage in these cells which is a consequence of exposure to cigarette smoke products.

2. Studies on the mutagenicity of urine from cigarette smokers

The original demonstration that cigarette smoke condensate, that is the particulate matter of cigarette smoke which we sometimes refer to as tar, was mutagenic, was made by Keir et al. (1974), using the *Salmonella typhimurium* system. These authors showed that the condensate from no more than 1 or 2% of the total tobacco in a single cigarette, could give significant increases in the frequencies of mutation measured as reversion to histidine prototrophy, if the condensates were incubated in the presence of S9 mix. No mutagenic activity was detected in the absence of the metabolising system. It was later shown (Hutton and Hackney, 1975) that most of this mutagenic activity was in the basic fraction of the smoke condensate, with no detectable mutagens being present in the neutral fractions — which are those that contain the polycyclic aromatic hydrocarbons such as benzo[a]pyrene. This is a point we shall again return to later. In the same year that Bruce Ames and his colleagues showed that cigarette smoke condensate was mutagenic to bacteria they (Durston and Ames, 1974), and also Barry Commoner et al. (1974), reported that Salmonella mutational assay system could also be used to detect mutagens present in the urines of rats fed carcinogens/mutagens.

It is these two findings, therefore, (i) that smoke condensates from cigarettes contain potent mutagens, and (ii) that it was possible to demonstrate the presence of mutagens in urine, that led to the study by Yamasaki and Ames (1977) on the mutagenicity of urine samples from cigarette smokers and non-smokers. These authors described a method for concentrating mutagens/carcinogens from human urine by a factor of around 200 by passing it through a non-polar XAD-2 amberlite resin. Material adsorbed to the polymer was eluted by acetone, the residue dried and then dissolved in dimethyl sulphoxide and assayed for mutagenicity using the *Salmonella typhimurium*/mammalian microsome test system. Yamasaki and Ames studied urines from 9 inhaling smokers, 2 non-inhaling smokers and 21 non-smokers, and showed that in the urine samples from smokers there were components that were mutagenic to Salmonella in the presence of S9 mix, and these components were absent, or present in only very small amounts, in non-smokers or in non-inhaling smokers. Moreover, the results showed that these mutagens were diminished in quantity in the urine of smokers if urine samples were taken following a period of abstinence from smoking, such as immediately following a night's rest, but were increased following a day of smoking activity, i.e. when urine was sampled on retiring in the evening.

These results clearly show that the urine of cigarette smokers contains mutagens which require activation by microsomes before being mutagenic to bacteria and they have been confirmed and extended in a number of laboratories including our own. One recent and particularly interesting confirmation is that described by Gelbart and Sontag (1980) in a study of smokers, non-smokers and alcoholic patients with cirrhosis of the liver. Using Salmonella strains TA98 and TA100 they showed that the urine from 11 of 12 cigarette smokers who smoked more than 10 cigarettes prior to urine collection, were positive in the mutagenicity tests and indeed they were positive in the absence of S9 mix when tested on TA100. 13 of 15 non-smoking controls had non-mutagenic urines and all 5 cirrhotics had strongly mutagenic urines (Fig. 2). The cirrhotics were all non-smokers and all had stopped drinking alcohol for more than 6 weeks prior to the test, and were on a routine hospital diet. Low protein diets are known to reduce the amount of cytochrome P450 and hence give less detoxification of carcinogens/mutagens, so that the mutagens in the urines from the cirrhotic patients may well be direct-acting mutagens that had come through in the diet of the patients.

In our studies on the mutagens in urine samples from smokers and non-smokers we have confirmed the results reported by Ames and his colleagues, but our main interest has been to see if we can isolate materials from smokers urine which give rise to chromosome aberrations and sister-chromatid exchanges in cells in culture. We are also looking at chromosome aberrations in cirrhotic patients and all this work is at a somewhat early stage. However, I can report that we are able to show that the urine of smokers but not non-smokers con-

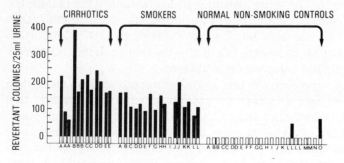

Fig. 2. Mutagenicity of urine from non-smoking cirrhotic patients, smokers and non-smoking controls. (After Gelbart and Sontag, 1980)

tains materials that, in the presence of S9 mix, give rise to a 50% increase in SCE frequencies in cultured cells of the Chinese hamster.

Now the two main conclusions we can draw from the work that I've referred to are: (i) that tobacco smoke condensate contains substances that are mutagenic to Salmonella, and (ii) that the urine of cigarette smokers, but not non-smokers, contains substances which are mutagenic in this bacterial system and substances which induce SCEs in mammalian cells in culture.

Following on these bacterial studies, and particularly during the last two years, a great deal of work has been undertaken to try and identify the mutagenic components in tobacco tars, and we shall return later to the nature of these mutagens.

3. Studies on chromosome damage in cells of cigarette smokers and in human cells exposed to smoke condensates in vitro

At around the same time that Yamasaki and Ames were investigating urine samples from smokers for mutagenicity, Obe and Herha (1978) in Berlin and my own group in Edinburgh were looking at the peripheral blood lymphocytes of smokers and non-smokers to see whether the smoking population had a higher frequency of chromosome aberrations in their blood cells. The Berlin population consisted of 20 individuals who were very heavy smokers, smoking from 40—60 cigarettes per day and having smoked for periods from 9 to 58 years and it was shown that the incidence of exchange aberrations was significantly higher in smokers. The Edinburgh population consisted of 50 males and 50 females randomly selected from a General Practice population without prior knowledge of their smoking habits. After completion of the scoring on this population it was ascertained that approximately half of these individuals were smokers although the vast majority smoked less than 20 cigarettes per day

TABLE 1

CHROMOSOME DAMAGE IN PERIPHERAL BLOOD LYMPHOCYTES IN EDINBURGH CIGARETTE SMOKERS

	Non-smokers	Smokers
Total subjects	43	55
Females	23	27
Males	20	28
Mean age (range)	45.8 (20—75)	48.2 (25—75)
Females	48.7 (20—75)	49.4 (25—75)
Males	43.0 (25—65)	47.0 (25—60)
Total metaphases analysed	4300	5500
Cells with damaged chromosomes (%)	109 (0.025)	176 (0.032)
Chromatid gaps	86 (2.006)	140 (2.545)
Chromatid breaks	24 (0.006)	19 (0.003)
Chromatid interchanges[a]	2 (0.047)	12 (0.218)
Chromosome gaps[a]	11 (0.256)	31 (0.564)
Chromosome breaks (fragments)[a]	14 (0.326)	37 (0.673)
Chromosome interchanges[a]		
asymmetrical	12 (0.0028)	23 (0.0042)
symmetrical	7 (0.0016)	11 (0.0020)

[a] Significant at 5% level.

and so we cannot classify them as heavy smokers. The results of the Edinburgh studies are summarised in Table 1 and again show an increased frequency of exchange aberrations in smokers relative to non-smoking controls. As we have already made the point that smoke products are distributed throughout the body, it may not be surprising then that the effects of these smoke-related mutagens can be observed in the form of chromosome damage in circulating blood lymphocytes, particularly in heavy smokers.

In addition to looking for the presence of gross chromosome rearrangements in peripheral blood cells of smokers, we in Edinburgh also looked to see if we could pick up any differences in the frequency of sister-chromatid exchange between smoking and non-smoking populations. In our studies on this first population we observed no difference in SCE incidence between cigarette smokers and non-smokers. On the other hand, Lambert et al. (1978) in Sweden reported a small, but significant, increase in SCE frequency in cells from relatively heavy cigarette smokers. At that time it seemed important to us to examine the reaction if cigarette smoke with the DNA of human cells exposed in vitro to see whether such exposure could indeed result in increased frequencies of SCEs. We therefore used a cigarette smoking machine to smoke 3 brands of cigarettes, high, medium and low tar and collected the condensate from each of these types of cigarette. The yields of condensate from smoke from the 3 cigarette brands are shown in Table 2 and these condensates were dissolved in DMSO and diluted with water to required concentrations before being applied to cultures of human lymphocytes containing 10^6 lymphocytes in 10 ml of culture medium. It was found that all condensates were potent inducers of SCE and that although the amounts of condensate produced by the high-, mid- and low-tar cigarettes differed, the equivalent amount of condensate from the 3 brands gave equal numbers of SCE (Fig. 3).

This kind of experiment has been repeated on a number of cigarette tars and on blood cells from a range of different people. In all cases the cigarette tars are potent inducers of SCE and, in general, we find that the SCE frequency is more than doubled by the addition of 0.5 mg of condensate to a culture (Fig. 4) i.e. a dose representing 1/80 of a high tar or 1/20 of a low tar cigarette — in other words in the smoke condensing from less than one puff of a cigarette!

Now it is known that there are some 1200 different compounds in cigarette smoke (Falk, 1977) and they include a range of precarcinogens, carcinogens,

TABLE 2

CIGARETTE-CONDENSATE YIELDS

Tar category of cigarette	Average dimensions of cigarette		Butt length (mm)	Weight of condensate (wet in mg) collected from 24 cigarettes smoked together[a]
	Length (mm)	Weight (mg)		
High	70	1008	20	1020
Middle	70	988	20	530
Low	83 (including 22 mm filter)	1012	25 (including filter)	270

[a] Each cigarette received in turn puffs of 35 ml lasting 2 sec at a frequency of once per minute until desired butt length reached.

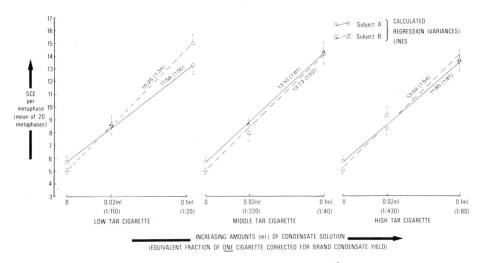

Fig. 3. The effect of 3 different cigarette-smoke condensates in producing SCEs in human lymphocytes cultured in vitro. 0.02 ml of condensate solution contains 0.1 mg of condensate dissolved in DMSO and diluted with water.

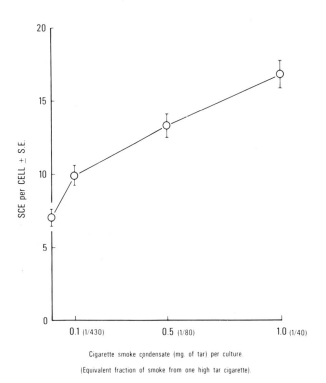

Fig. 4. Relation between dose of tar condensate and SCE induction in human lymphocytes exposed to various concentrations of condensate in vitro.

procarcinogens and promotors and, at least up until a year or so ago, the polycyclic aromatic hydrocarbons, and in particular the eventual metabolite or metabolites of benzo[a]pyrene (BP), were regarded as the most potent carcinogens. We therefore looked at the potency of benzo[a]pyrene in inducing SCEs in our human lymphocyte test system (Fig. 5) and found that some 2.5 µg of benzo[a]pyrene, which is an amount equal to that found in the smoke of 40—200 cigarettes, is required to produce similar 2-fold increases in SCEs to 0.5 mg of smoke condensate: that is 1/80th of a cigarette. In other words, weight for weight, cigarette smoke condensate is between 4000 and 20000 times more potent than the benzo[a]pyrene contained within it, in inducing SCEs in human lymphocytes in vitro. Our conclusion that benzo[a]pyrene was not responsible for the increased SCE in human lymphocytes exposed to tobacco smoke condensate was further reinforced by some studies that we carried out in Chinese hamster cells (Hopkin and Perry, 1980).

Chinese hamster cells are relatively insensitive to induction of SCEs by compounds that require microsomal activation such as benzo[a]pyrene. However, if we add S9 then benzo[a]pyrene is a potent SCE inducer, but its activation can be blocked if α-naphthoflavone (ANF) is present in the culture medium. In contrast, cigarette smoke condensate is potent in inducing SCEs in the presence or absence of S9 and when we add ANF to cigarette smoke condensate this has no effect whatsoever in minimising the effects of condensate in inducing SCEs (Fig. 6). These results on Chinese hamster cells which show, (i) that smoke condensate is active in the absence of S9 mix and (ii) that its action is not diminished in the presence of α-naphtho-flavone, clearly indicate that the induction of SCEs is not due to benzo[a]pyrene or indeed to any other polycyclic hydrocarbon that requires activation by the microsome system.

These findings parallel some of the earlier work that showed that polycyclic

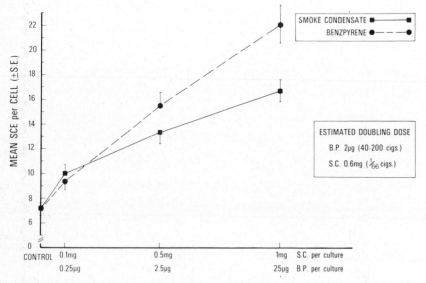

Fig. 5. Effects of tobacco-smoke condensate (S.C.) and benzo[a]pyrene (B.P.) in inducing SCE in cultured human lymphocytes. The estimated doubling doses are for B.P. 2 µg and S.C. 0.6 mg/culture.

aromatic hydrocarbons present in very small quantities in cigarette smoke condensate could not by themselves account for the carcinogenic effect of condensate. Moreover, the recent studies by Matsumoto et al. (1977, 1978) and by Nagao et al. (1977a, b) in Japan have shown that the important mutagenic principles in tobacco tars and in burned organic materials are most probably the aromatic amines that are pyrolysis products of amino acids and proteins. In particular this work has shown that the major active substances that are mutagenic in Salmonella are two pyrolysates from tryptophan labelled TRP-P-1 and TRP-P-2 and one from the glutamic acid, GLU-P-1 (Fig. 7). Tohda et al. (1980) have further shown that treatment of Syrian hamster fibroblasts in vitro with these three compounds results in their transformation and such transformed cells when transferred to hamster cheek pouches result in tumours. Moreover, the injection of these compounds into Syrian hamsters results in malignant fibrosarcomas at the sites of injection, so that these mutagenic substances from the pyrolysates of amino acids and proteins are therefore both carcinogens and mutagens. Tohda et al. (1980) have also examined the effects of these 3 pyrolysates in inducing SCEs in a line of human lymphoblastoid cells and showed that they were active in inducing SCEs (Fig. 8). Comparative studies on their potency relative to other known SCE inducers show that TRP-P-2 is weight for weight as potent as aflatoxin B_1 and is some 3 times as effective as TRP-P-1, which is itself at least as effective as benzo[a]pyrene. However, we should note that SCE induction by these pyrolysis products necessitates their activation by S9, whereas we have shown that the ingredients

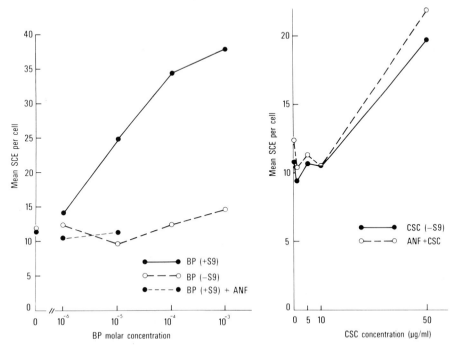

Fig. 6. SCE induction in Chinese hamster cells exposed to benzo[a]pyrene (B.P.) and cigarette-smoke condensate (C.S.C.) in the presence or absence of S9. Note the requirement for S9 to activate B.P. and its inactivation in the presence of α-naphthoflavone (ANF), whereas the action of CSC is independent of both S9 and ANF.

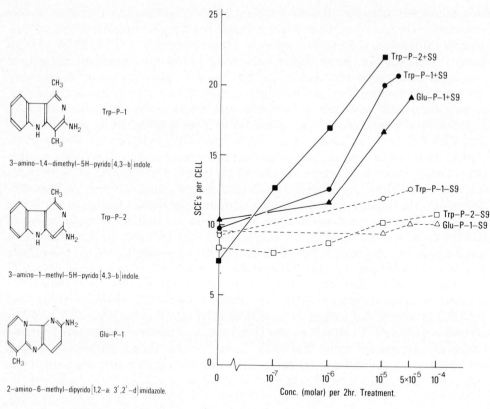

Fig. 7. Mutagenic aromatic amines from pyrolysates of tryptophan and glutamic acid.

Fig. 8. The induction of SCEs in human lymphoblastoid cells by pyrolysates of tryptophan (Trp-P-1 and 2) and glutamic acid (Glu-P-1). Note that these compounds are inactive in the absence of S9. (Data after Tohda et al., 1980)

in cigarette tobacco tar that induce SCEs in human lymphocytes or Chinese hamster cells are direct-acting and are not dependent upon the presence of mixed function oxidases.

In addition to the discoveries of the mutagenicity, carcinogenicity, and SCE-inducing properties of the pyrolysis products of proteins, earlier studies by Matsumoto et al. (1977) had shown that pyrolysis of proteins results in the formation of two substances referred to as harman and norharman, which are not in themselves carcinogenic, but act as tumour promotors and also markedly enhance the mutagenicity of the active ingredients in tobacco condensates. I previously referred to the fact that there may be a synergistic action between cigarette smoking and other environmental factors in producing lung cancer and I should simply point out here that the action of many of the active ingredients in cigarette smoke can be enhanced by the presence of other components. Similarly, and in direct contrast, other compounds may interact with active agents to nullify their effects, and here it is of interest to note that nitrites, which may themselves yield carcinogenic N-nitroso compounds, will deaminate and completely inactivate the mutagenic aromatic amines at low pH (Yoshida and Matsumoto, 1978; Tsuda et al., 1980); whereas increasing nitrates in tobacco (by using increased amounts of nitrogenous fertilizer) yields tars with increasing mutagenicity (Mizusaki et al., 1977). Tobacco-tar chemistry is complex and is not something that I would wish to pursue. Instead, I should like to discuss some other work in our laboratory which arose from our finding that SCE induction was not a consequence of the presence of benzo[a]-pyrene.

A considerable amount of activity was stimulated by the original work of Kellermann et al. (1973a, b) in their report that different individuals had different degrees of inducibility for the enzyme complex aryl hydrocarbon hydroxylase (AHH). This AHH complex is involved in the metabolism and activation of precarcinogenic polycyclic hydrocarbons such as benzo[a]pyrene and the data of Kellermann et al. (1973a, b) indicated that populations of individuals with lung cancer appeared to contain a much higher incidence of high level inducers than controls without the disease. There is some evidence for a genetic factor in lung cancer (Tokuhata and Lilienfeld, 1963) and the inducibility of AHH in the mouse is controlled by a single genetic locus (Nebert et al., 1972) but in man may involve many more loci (Coomes et al., 1976) and, indeed, the original clear association between AHH inducibility and cigarette-smoke-induced lung cancer in man has, by and large, not been convincingly confirmed (Paigen et al., 1977; Lieberman, 1978). Our results on SCE induction by tobacco condensate suggested to us that perhaps looking for differences between the responses of cells from different individuals to the activation of polycyclic aromatic hydrocarbons might be less relevant than looking at the response to SCE induction in vitro by tobacco-tar condensates in cells from different individuals. We therefore (Hopkin and Evans, 1980) set up a study to look at the response of lymphocyte chromosomes from 4 different groups of people:

(A) 10 healthy cigarette smokers, none of whom to our knowledge had been unduly exposed to any mutagen other than cigarette smoke.

(B) 10 healthy non-smokers who were examined in pairs with the people in group (A) and were matched with them on the basis of age and sex.

(C) 12 patients with histologically proven, but untreated, non-disseminated anaplastic or squamous cell bronchial carcinomas, in whom diagnostic biopsies had been obtained by fiberoptic bronchoscopy. 10 of these 12 patients were continuing to smoke cigarettes at the time of testing and these were labelled C1, but 2 had stopped smoking 2 and 5 years previously and these were labelled C2.

(D) Our fourth and final group consisted of 10 control patients who were selected as controls for Group C. None of the group D controls had any form of cancer and they were on average 10 years older and had smoked a similar number of cigarettes per day, but for 10 years longer, than our lung cancer patients. All the people in groups (D) and (C) were inhaling smokers and smoked their cigarettes to comparable butt lengths.

Blood lymphocytes from each of these individuals in the 4 groups were cultured both in the absence and presence of various concentrations of tobacco-tar condensate and dose-response curves for SCE induction obtained for each individual. The results from the healthy smokers and healthy non-smokers showed typical dose-response curves for SCEs against cigarette-smoke condensate in mg per culture (Fig. 9), and although there is considerable scatter, and each point represents a determination from a single individual, it is pretty evident that the SCE levels in smokers are higher than in non-smokers. This difference is true for the basal levels, that is the levels observed in blood cells not exposed to tar in culture, as well as in the levels observed following exposure to tar condensate in vitro. The results from the lung-cancer patients and their matched controls again show a reasonable scatter (Fig. 10), but it seems pretty clear that the lung-cancer subjects show a higher incidence, at least in their response to smoke condensate in culture, than the smoking matched controls.

The pooled data for each of the groups are summarised in Fig. 11 and this shows that the basal SCE frequencies of the smokers, i.e. the 3 groups (B), (C) and (D), are not different, but are higher than the basal SCE frequencies in healthy non-smokers. The almost parallel dose-response curves following exposure to condensate in vitro might imply that the increased SCE rates following in vitro induction may simply emphasise in vivo differences, with

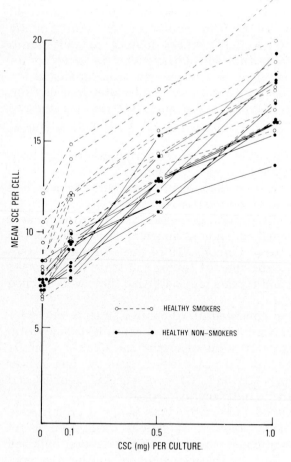

Fig. 9. Mean SCE frequencies (20 cells per point) in lymphocytes of healthy non-smokers and healthy smokers exposed to cigarette smoke condensate in vitro.

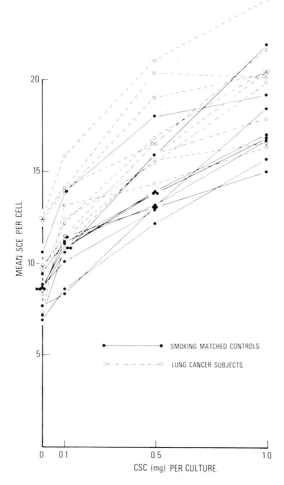

Fig. 10. Mean SCE frequencies (20 cells per point) in lymphocytes of lung-cancer subjects and older smoking matched controls exposed to cigarette-smoke condensate in vitro.

similar increments being added to all 4 groups for each in vitro unit dose. This interpretation cannot be excluded, but at lower doses of concentrate the early response is certainly not uniform between groups. What is clearly obvious is that the in vitro induction highlights the differences between the 4 groups. Highest responses are found for smokers with lung cancer (C1). Amongst smokers the lowest response was found in the older heavy smoking non-cancer patients group (D) and this is in keeping with our selection of these people as possibly representing a relatively low risk group. An unselected group of healthy and younger smokers (A) who might reasonably be expected to contain individuals of varying risk for lung cancer, show a mean response intermediate between the lung-cancer subjects (C1) and their controls (D). This, then, demonstrates an association of risk with the extent of DNA damage and is in line with the possibility that somatic mutation might be important in initiating the malignant transformation in cells.

Our results also show that the number and type of cigarettes smoked

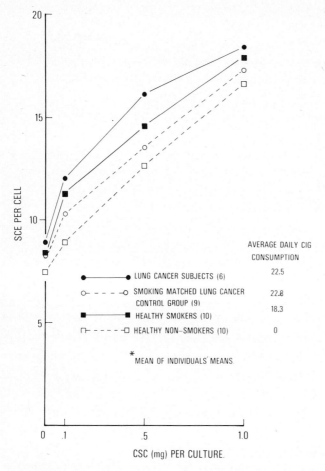

Fig. 11. Pooled "within group" data on SCE frequencies in lymphocytes of non-smokers, smokers and lung-cancer patients exposed to cigarette-smoke condensate in vitro.

significantly affect SCE rates, the heaviest smokers show the highest frequencies and it is clear that smokers seemingly exposed to the same amounts of cigarette smoke in vivo can have quite different SCE rates as is evident as between groups (C1) and (D). These differences could result from discrepancies in the actual amounts of tobacco smoked and certainly all smoking histories describe only an alleged amount. Or they may result from varying patterns of inhalation or of air flow into the bronchial tree, both of which would clearly influence the dose of carcinogen delivered to the bronchial mucosa and to the peripheral lymphocytes. Alternatively, if real smoke dosage in vivo was equivalent in both groups, and there is no way of determining this precisely, then these differences might reflect the influence of other factors on the extent of measured DNA damage. In this context we might say that Tokuhata and Lilienfeld (1963) have shown that there is a significant familial aggregation of lung cancer which does not extend to spouses, so that genetic factors would appear to be involved.

Finally, I should say that the 3 sets of information I have presented are all in line with the notion that cigarette-smoke products interact with DNA and that cigarette smoke is mutagenic, and there are clear indications that mutagenicity is simply not confined to bacteria, but is relevant to man himself. On the basis of these data we would certainly expect that there must be heritable effects demonstrable in man as a consequence of cigarette smoking. Unfortunately, there are only a few limited studies and aside from the as yet unsubstantiated data of Mau and Netter (1974), which indicates a small but significant increase in perinatal mortality and congenital abnormalities in the new born when husbands smoke more than 10 cigarettes per day, we really have no firm evidence that exposing ourselves to tobacco smoke increases our mutation rate. We would nevertheless be both foolish and complacent to believe that it did not.

References

Bedford, J.M., M.J. Bent and H. Calvin (1973) Variations in the structural character and stability of the nuclear chromatin in morphologically normal human spermatozoa, J. Reprod. Fertil., 33, 19—29.

Commoner, B., A.J. Vithayathil and U.K. Henry (1974) Detection of metabolic carcinogen intermediates in urine of carcinogen-fed rats by means of bacterial mutagenesis, Nature (London), 249, 850—852.

Coomes, M.L., W.A. Mason, I.E. Muijsson, E.T. Cantrell, D.E. Anderson and D.L. Busbee (1976) Aryl hydrocarbon hydroxylase and 16α-hydroxylase in cultured human lymphocytes, Biochem. Genet., 14, 671—685.

Doll, R. (1977) Introduction, in: Origins of Human Cancer, Book A, Cold Spring Harbor Conference, pp. 1—12.

Doll, R., and R. Peto, (1978) Cigarette smoking and bronchial carcinoma: dose and time relationships among regular smokers and lifelong nonsmokers, J. Epidemiol. Community Hlth. 32, 303—313.

Dontenwill, W., H.-J. Chevalier, H.-P. Harke, U. Lafrenz, G. Reckzeh and B. Schneider (1973) Investigations on the effects of chronic cigarette-smoke inhalation in Syrian golden hamsters, J. Natl. Cancer Inst., 51, 1781—1832.

Durston, W.E., and B.N. Ames (1974) A simple method for the detection of mutagens in urine: studies with the carcinogen 2-acetylaminofluorene, Proc. Natl. Acad. Sci. (U.S.A.), 71, 737—741.

Evans, H.J., J. Fletcher, M. Torrance and T.B. Hargraeve (1981) Sperm abnormalities and cigarette smoking, Lancet, 1, 627—629.

Falk, H.L. (1977) Chemical agents in cigarette smoke, in: D.H.K. Lee (Ed.), Handbook of Physiology, Reaction to Environmental Agents, Waverley, Baltimore, pp. 199—211.

Gelbart, S.M., and S.J. Sontag (1980) Mutagenic urine in cirrhosis, Lancet, i, 894—896.

Hammond, E.C., and I.J. Selikoff (1973) Relation of cigarette smoking to risk of death of asbestos-associated diseases among insulation workers in the United States, in: Biological Effects of Asbestos, IARC Publication No. 8, Int. Agency Res. on Cancer, Lyon, pp. 312—317.

Hopkin, J.M., and H.J. Evans (1980) Cigarette smoke-induced DNA damage and lung cancer risks, Nature (London), 283, 388—390.

Hopkin, J.M., and P.E. Perry (1980) Benzo[a]pyrene does not contribute to the SCEs induced by cigarette smoke condensate, Mutation Res., 77, 377—381.

Hutton, J.J., and C. Hackney (1975) Metabolism of cigarette smoke condensates by human and rat homogenates to form mutagens detectable by *Salmonella typhimurium* TA1538, Cancer Res., 35, 2461—2468.

ICPEMC Publication No. 3 (1979) Cigarette smoking — does it carry a genetic risk? Mutation Res., 65, 71—81.

Kier, L.D., E. Yamasaki and B.N. Ames (1974) Detection of mutagenic activity in cigarette smoke condensates, Proc. Natl. Acad. Sci. (U.S.A.), 71, 4159—4163.

Kellermann, G., M. Luyten-Kellerman and C.R. Shaw (1973a) Genetic variation of aryl hydrocarbon hydroxylase in human lymphocytes, Am. J. Hum. Genet., 25, 327—331.

Kellermann, G., C.R. Shaw and M. Luyten-Kellerman (1973b) Aryl hydrocarbon hydroxylase inducibility and bronchogenic carcinoma, N. Engl. J. Med., 289, 934—937.

Lambert, B., A. Lindblad, M. Nordenskjöld and T.B. Werelius (1978) Increased frequency of sister chromatid exchanges in cigarette smokers, Hereditas, 88, 147—149.

Lieberman, J. (1978) Aryl hydrocarbon hydroxylase in bronchogenic carcinoma, N. Engl. J. Med., 298, 686—687.

Matsumoto, T., D. Yoshida and S. Mizusaki (1977) Enhancing effect of harman on mutagenicity in Salmonella, Mutation Res., 56, 85—88.

Matsumoto, T., D. Yoshida, S. Mizusaki and H. Okamoto (1977) Mutagenic activity of amino acid pyrolysates in *Salmonella typhimurium* TA98, Mutation Res., 48, 279—286.

Matsumoto, T., D. Yoshida, S. Mizusaki and H. Okamoto (1978) Mutagenicities of the pyrolyzates of peptides and proteins, Mutation Res., 56, 281—288.

Mau, G., and P. Netter (1974) Auswirkungen des väterlichen Zigarettenkonsums auf die perinatale Sterblichkeit und Missbildungshäufigkeit, Dtsch. Med. Wschr., 99, 1113—1118.

Mizusaki, S., T. Takashima and K. Tomaru (1977) Factors affecting mutagenic activity of cigarette smoke condensate in *Salmonella typhimurium* TA1538, Mutation Res., 48, 29—36.

Nagao, M., M. Honda, Y. Seino, T. Yahagi and T. Sugimura (1977a) Mutagenicities of smoke condensates and the charred surface of fish and meat, Cancer Lett., 2, 221—226.

Nagao, M., M. Honda, Y. Seino, T. Yahagi, T. Kawachi and T. Sugimura (1977b) Mutagenicities of protein pyrolysates, Cancer Lett., 2, 335—340.

Nebert, D.W., F.M. Goujon and J.E. Gielen (1972) Aryl hydrocarbon hydroxylase induction by polycyclic hydrocarbons: simple autosomal dominant trait in the mouse, Nature (London), New Biol., 236, 107—110.

Oakberg, E.F., and R.L. DiMinno (1960) X-Ray sensitivity of primary spermatocytes of the mouse, Int. J. Radiat. Biol., 2, 196—209.

Obe, G. (1981) These proceedings, pp. 19—23.

Obe, G., and J. Herha (1978) Chromosomal aberrations in heavy smokers, Hum. Genet., 41, 259—263.

Paigen, B., H.L. Gurtoo, J. Minowada, L. Houten, R. Vincent, K. Paigen, N.B. Parker, E. Ward and N.T. Hayner (1977) Questionable relation of aryl hydrocarbon hydroxylase to lung-cancer risk, N. Engl. J. Med., 297, 346—350.

Royal College of Physicians (1977) Smoking and cancer, in: Smoking or Health, Pitman, London, pp. 52—64.

Seuanez, H.N., A.D. Carothers, D.E. Martin and R.V. Short (1977) Morphological abnormalities in spermatozoa of man and great apes, Nature (London), 270, 345—347.

Tohda, H., A. Oikawa, T. Kawachi and T. Sugimura (1980) Induction of sister-chromatid exchanges by mutagens from amino acid and protein pyrolysates, Mutation Res., 77, 65—69.

Tokuhata, G.K., and A.M. Lilienfeld (1963) Familial aggregation of lung cancer in humans, J. Natl. Cancer Inst., 30, 289—312.

Tsuda, M., Y. Takahashi, M. Nagao, T. Hirayama and T. Sugimura (1980) Inactivation of mutagens from pyrolysates of tryptophan and glutamic acid by nitrite in acidic solution, Mutation Res., 78, 331—339.

Viczian, M. (1969) Ergebnisse von Spermauntersuchungen bei Zigarettenrauchern, Z. Haut-Geschl.-Krkh., 44, 183—187.

White, J.R., and H.J. Froeb (1980) Small-airways dysfunction in nonsmokers chronically exposed to tobacco smoke, New Engl. J. Med., 302, 720—723.

WHO Technical Report Series (1975) No. 568, Smoking and its Effects on Health.

Wyrobek, A.J., and W.R. Bruce (1975) Chemical induction of sperm abnormalities in mice, Proc. Natl. Acad. Sci. (U.S.A.), 72, 4425—4429.

Yamasaki, E., and B.N. Ames (1977) Concentration of mutagens from urine by adsorption with the nonpolar resin XAD-2: cigarette smokers have mutagenic urine, Proc. Natl. Acad. Sci. (U.S.A.), 74, 3555—3559.

Yoshida, D., and T. Matsumoto (1978) Changes in mutagenicity of protein pyrolyzates by reaction with nitrite, Mutation Res., 58, 35—40.

HOW RELEVANT ARE HIGH DOSES IN MUTAGENICITY AND CARCINOGENICITY STUDIES IN ANIMALS?

H. GREIM, U. ANDRAE, W. GÖGGELMANN, S. HESSE, L.R. SCHWARZ and K.H. SUMMER

Department of Toxicology, Gesellschaft für Strahlen- und Umweltforschung, D-8042 Neuherberg (F.R.G.)

Summary

Toxicity, including genotoxicity of chemicals, is frequently affected by metabolic conversion of the chemicals by the host, leading to inactive derivatives. Inactivation mechanisms can be overcome by high doses of a chemical resulting in a greatly altered dose—effect relationship and drastically increased toxicity. This is exemplified by the dose-dependent effect of chloroform and 1,1-dichloroethylene in animals, by investigations on the glutathione-dependent mutagenicity of chlorodinitrobenzene in the Ames test, and by the effects resulting from inhibited metabolic inactivation of naphthalene in isolated hepatocytes.

At low doses genotoxic effects of chemicals may be prevented by metabolic inactivation, e.g. chloroprene is mutagenic in vitro but not carcinogenic, possibly due to glutathione conjugation. Moreover, disputable data on mutagenicity or carcinogenicity of styrene must be judged in that its mutagenic metabolite styrene oxide is inactivated by efficient cellular epoxide hydrolase activity in vivo.

Hepatic monoxygenase activity creates mutagenic hydrogen peroxide which is trapped in vivo by catalase and glutathione peroxidase. Thus, information on the mechanism and extent of inactivation is essential for extrapolation from high doses of chemicals used in toxicity studies in animals to low doses that are relevant to human exposure.

Introduction

For risk estimation of man's exposure to carcinogenic chemicals it is customary to extrapolate from high doses at which tumor induction is observed in animals to low doses in the human environment. However, there is increasing evidence that several carcinogens induce other biological responses at low doses than those at high doses. Furthermore, species differences in the metabolic activation and inactivation processes may also question a linear dose—effect relationship.

An evaluation of these facts is essential because one has to be aware that man is continuously exposed to small amounts of mutagens and carcinogens. Commoner et al. (1978) reported the presence of mutagens in broiled hamburgers; the presence of mutagenic activity in charred fish and beef was previously reported by Nagao et al. (1977) and Sugimura et al. (1977). Such mutagens apparently stem from pyrolisation of the amino acids tryptophan and glutamic acid (Sugimura, 1978). Nitrosamines present in food (Spiegelhalder et al., 1979; Scanlan et al., 1980) are constantly formed in the human stomach when secondary and tertiary amines are ingested together with nitrite (Lijinsky, 1977; Tannenbaum, 1980). Chloroform has been used for many years in tooth pastes and in cough syrup formulations and it is a drinking-water contaminant whenever chlorine is added for hygienic reasons (Fishbein, 1979; Stockinger, 1977). Sugimura recently evaluated that by the age of 50 a person has eaten approx. 10 tons dry weight of food (Sugimura, 1978). Even though concentrations of environmental contaminants in food are very low, their continuous intake could be sufficient to cause genetic damage possibly resulting in the appearance of cancer.

Models to extrapolate from high to low doses and between species

Animal studies to investigate the potential hazards of such chemicals are usually performed at relatively high levels of exposure (Crump et al., 1976; Cornfield, 1977; Scientific Committee, 1978; Maugh, 1978; Tomatis, 1979; Carter, 1979; Interagency Regulatory Liaison Group, 1979). In general, human risk is determined by extrapolating from the high doses used in the animal experiments to the much lower levels to which man is actually exposed assuming that no threshold exists. This is based on the concept that cancer is induced after the reaction of a single or several molecules with DNA or other macromolecules in a cell. Once such a reaction has occurred the cell is damaged irreversibly and initiates a new disorganized cell line which finally produces the tumor. This mechanism implies that there will be a certain tumor incidence in a population no matter how long the dose of the carcinogen has been. In this case, cancer incidence at low doses will be a linear function of the high doses.

The formula generally applied in extrapolation from high to low doses is (Reitz et al., 1978): Linear $P = \beta \times d$ where P is the fractional incidence of excess tumors, d the average daily dose in mg/kg and β is a dimensionless parameter to be estimated from the experimental data. Additionally, a species correction factor can be applied to β which has been chosen as the cube root of the ratio of the body weights to calculate the dose per surface area. This always implies that man is considered to be more sensitive to a carcinogen than mice or rats. Rall (1960) established this correlation with 18 anti-neoplastic agents. They produced, however, direct toxicity, and Rall stated that these compounds are not involved in the various drug-metabolizing systems. However, there is ample evidence that most of the environmental chemicals undergo metabolism by the host, and it becomes evident from this that the rate of metabolism must be included into the process of risk estimation.

Reitz et al. (1978) suggested therefore that the relative rates of production of toxic metabolites be introduced into risk estimation. And, indeed, relative carcinogenicity of chloroform, vinylidene chloride, 2-acetylaminofluorene and trichloroethylene in rats and mice correlated well with the relative rate of metabolic activation to a reactive intermediate of each of these chemicals.

The resolution of these approaches by animal experiments seems to be impossible. To decide which of the correction factors is correct would mean that data on the response to low doses of the chemical concerned would have to be collected. But this would mean measuring low incidences of cancer in animal populations, say 0—15%, and these low incidences can only be distinguished with any degree of statistical certainty from the natural rate of cancer in animals by trials involving impossibly large numbers of both control and treated animals. Thus, in practice, only relatively high doses can yield statistically significant data. But in many cases such high doses may produce cancer simply because their very quantity overwhelms the biochemical pathways that would detoxify smaller, more realistic doses.

Since no conclusive solution of the problem can be obtained from the standard carcinogenicity or mutagenicity test procedures the only approach to solve the problem should be the evaluation of mechanisms which may affect the tumor incidence at low and high doses of carcinogens. This may provide a more solid basis for the presently highly speculative and uncertain quantitative risk assessment.

In general several mechanisms are considered to be involved in the inactivation of chemical carcinogens. Important contributors to these mechanisms are membranes that may not permit potential carcinogens to come into contact with DNA and enzyme systems for repair of damage to DNA. Some evidence also suggests that the immune system can identify and destroy many kinds of operant cells produced by chemicals that slip through the other defenses.

Moreover, enzyme systems for metabolic toxification and detoxification modify the biological impact of chemicals and have to be considered in the course of risk estimation.

This is explained in more detail by discussing the following points:

(1) Is there any indication of threshold doses of carcinogenic chemicals in man due to metabolism?

(2) Are the high doses of carcinogens used in animal studies generally representative for the low doses to which man is exposed?

(3) Does metabolism result in qualitative and quantitative differences in the susceptibility of different species to carcinogens?

Indication of threshold doses due to metabolic inactivation

(a) Carcinogenicity of estrone

The hormone estrone, for example, has been shown by several investigators to be carcinogenic when given to laboratory animals in large doses (IACR, 1974). However estrone is present in very small concentrations in all humans, yet without demonstrable evidence of harm. It is present in much higher concentrations in women than in men which might be at least part of the cause of the much greater incidence of breast cancer in women.

Moreover, glucuronidation of estrogens seems to be a significant inactivating mechanism. A recent report by Fishman et al. (1979) indicates a correlation between low glucuronidation capacity and the occurrence of breast cancer in women. These authors studied steroid excretion in the urine of 30 young women who by genetic analysis were at risk for familial breast cancer as compared to a carefully matched control group. Highly significant differences were observed in the excretion of glucuronides of estrone and estradiol with the high-risk women exhibiting lower values than the controls. Besides the finding of a potential discriminant for identifying women at high risk for breast cancer, this observation strongly indicates that the status of inactivation mechanisms affects susceptibility to potential carcinogens.

(b) Increased DNA-repair by reactive oxygen species

Recently, we demonstrated that reactive oxygen species, which are formed during monooxygenase-mediated metabolism of chemicals, induce DNA-repair activity in the human lymphoblastoid cell line NC37 BaEV (Andrae and Greim, 1979). This was observed in the course of measuring the induction of DNA-repair replication by mutagens requiring metabolic activation (Fig. 1). Muta-

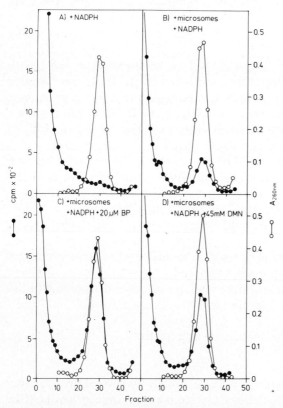

Fig. 1. Repair replication in NC37 BaEV cells. Cultures were incubated in the presence of 1 μM 5-fluorodeoxyuridine and 10 μM 5-bromodeoxyuridine for 1 h. After addition of 2.5 mM hydroxyurea cells were incubated for 3 h with A: NADPH; B: 0.5 mg protein/ml liver microsomes of clophen A 50 pretreated rats + NADPH; C: liver microsomes + NADPH + 20 μM benzo[a]pyrene; D: liver microsomes + NADPH + 45 min mM dimethylnitrosamine, and labeled for 4.5 h with 10 μCi [^{3}H]thymidine (40Ci/mmole). Cells were lysed in 0.5% sodiumdodecylhydrogensulfate, digested with proteinase K (50 μg/ml) and sedimented in alkaline CsCl-gradients (Andrae and Greim, 1979). The location of [^{3}H]-label at normal density (coincident with the absorbance peak) is indicative for repair replication.

gens such as benzo[a]pyrene and dimethylnitrosamine induce repair activity. However, in control experiments without the mutagen a marked incorporation of repair label was observed. The incubate was comprised of washed microsomes of phenobarbital-treated mice, NADPH and hydroxyurea which is used to suppress semiconservative DNA synthesis. When this inhibitor was omitted, no repair replication occurred. Hydroxyurea-induced repair was dependent on the presence of microsomes and NADPH, and was reduced to 50% by the metabolic inhibitor SKF 525-A. This highly indicates that hydroxyurea-induced repair is a consequence of microsomal monooxygenase-mediated metabolism. Since addition of glutathione (GSH) and superoxide dismutase reduced and catalase prevented hydroxyurea-induced DNA repair, hydrogen peroxide (H_2O_2) or reactive oxygen species deriving from H_2O_2 are likely to be the genotoxic agents. As H_2O_2 production generally occurs during monooxygenase function (Hildebrandt and Roots, 1975; Jones et al., 1978) we investigated the effect of ethylmorphine. Accordingly, ethylmorphine induced DNA repair in the test system but the effect was prevented in the presence of catalase (Fig. 2). Other in vitro cell systems, e.g. human A549 lung tumor cells, apparently have a sufficient H_2O_2-inactivation capability as these systems did not show enhanced repair activity.

The precise mechanisms by which reactive oxygen species are trapped in vivo remain unknown. GSH peroxidase is likely to be involved. It is evident that efficient H_2O_2-inactivation must be present in vivo to prevent reactive oxygen species from exceeding a no-effect concentration and thus preventing genotoxic effects.

(c) Chloroprene: Mutagenicity but no carcinogenicity

Further evidence for no-effect levels due to metabolic inactivation is given by toxicity studies on chloroprene (2-chloro-1,3-butadiene).

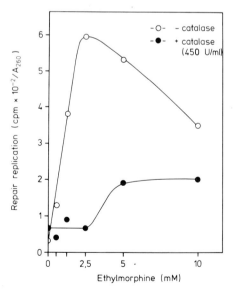

Fig. 2. Repair replication in NC37 BaEV cells in the presence of liver microsomes isolated from phenobarbital-treated mice, NADPH and ethylmorphine. For details see legend to Fig. 1.

Chloroprene is a reactive chemical which is widely used in the manufacturing of the synthetic rubber neoprene. Chloroprene has been suggested to be responsible for the increased incidence of skin and lung cancer in workers exposed to the chemical. However, the data are questionable and no carcinogenic effects of chloroprene have been noted in animal studies to date (Fishbein, 1979; Ponemarkov and Tomatis, 1980). Bartsch et al. (1975, 1979) have demonstrated a slight mutagenicity of chloroprene in *Salmonella typhimurium* strains without metabolic activation. Mutagenicity was increased about 3-fold in the presence of S9 fractions. These observations were considered to reflect the probable formation of an epoxide intermediate of chloroprene.

Why was chloroprene shown to be mutagenic but not carcinogenic? In analogy to vinylchloride and 1,1-dichloroethylene, a subsequent detoxification of the chloroprene by conjugation with glutathione has been suggested (Haley, 1978; Plugge and Jaeger, 1979). Since in vitro mutagenicity test procedures do not include sufficient glutathione (Summer et al., 1980), the apparent discrepancy of mutagenicity in vitro without producing carcinogenicity in the animal may be explainable. We tested the possible involvement of glutathione in inactivation in isolated rat hepatocytes and in the whole animal (Summer and Greim, 1980).

Similar to the rat liver, cellular glutathione decreased in isolated rat hepatocytes to about 50% of the control values within 15 min in the presence of 3 mM chloroprene (Fig. 3). This depletion was concentration-dependent and increased with time. In hepatocytes of animals pretreated with phenobarbital or Clophen A50, 3 mM of chloroprene almost completely depleted glutathione within 30 min.

This strongly indicates that a Phase I reaction, presumably epoxide for-

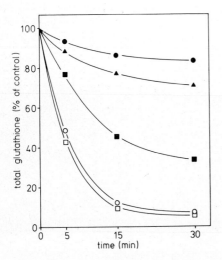

Fig. 3. Glutathione depletion in isolated rat hepatocytes by chloroprene. Data are from 1 representative experiment out of 3 with different cell preparations. Control level of cellular GSH amounted to 19.9 ± 2.1 nmoles/mg cell protein. Closed symbols: Hepatocytes (8 mg cell protein/ml) of untreated animals incubated with 0.5 mM (●), 1.0 mM (▲) and 3.0 mM (■) chloroprene. Open symbols: Hepatocytes from either phenobarbital- (○) or Clophen A 50- (□) pretreated animals incubated with 3 mM chloroprene.

mation which is enhanced after Clophen A50- or phenobarbital-pretreatment, precedes glutathione conjugation.

Glutathione-dependent detoxification in the animal was further verified by determining urinary thioether excretion comprised of GSH conjugates and mercapturic acids. Chloroprene administration to rats resulted in a dose-dependent increase in the excretion of urinary thioethers (Fig. 4). This increase was reversible and completed within 24 h at all dose levels administered. At dose levels of 50 and 100 mg chloroprene/kg, the additional excretion of urinary thioethers was almost 200 and 450 μmoles/kg daily, respectively.

It is to be noted that no linear dose—response relationship was observed in thioether excretion, indicating that at higher doses of chloroprene the availability of cellular glutathione becomes rate-limiting.

Thus, a dose level at which high concentrations of chloroprene overcome GSH-inactivation is suggested. Only sufficiently high doses are expected to induce carcinogenicity. They have not been used so far in carcinogenicity studies.

Effects of high doses on metabolism

(a) Overwhelmed metabolic inactivation

The latter observation suggests that high doses of a chemical may have other biological effects than lower doses by overcoming metabolic inactivation mechanisms.

Vinylidene chloride (1,1-dichloroethylene) is also conjugated to glutathione. In the rat, glutathione is depleted by exposure to high levels of 1,1-dichloroethylene (Jaeger et al., 1974). Fasting of rats prior to 1,1-dichloroethylene exposure further depleted the glutathione and, at the same time, dramatically increased the hepatotoxicity of the chemical. Furthermore, McKenna et al. (1977) reported that a sudden disproportionate increase in macromolecular binding of 1,1-dichloroethylene metabolites occurred when glutathione was depleted by more than 30%. These observations are strongly in favour of the protective function of such inactivating systems.

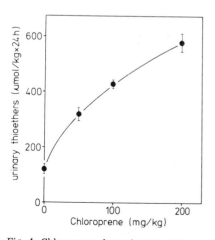

Fig. 4. Chloroprene-dependent excretion of thioethers in the urine of rats. Chloroprene was administered in olive oil by stomach tube. Data represent means ± S.D. from 4 animals.

(1) Chlorodinitrobenzene mutagenicity and GSH conjugation

This is further demonstrated by in vitro experiments using the plate incorporation assay according to Ames which consists of *S. thyphimurium* as target cells and the S9 fraction. The latter is a cell homogenate consisting of cytoplasma and microsomes without cell-membrane, mitochondria and nuclei.

Chlorodinitrobenzene is a direct mutagen with a dose-dependent increase in mutagenicity in the Salmonella test when added directly to the bacteria (Summer and Göggelmann, 1980). In the presence of postmitochondrial supernatant, the shape of the dose—effect curve was different (Fig. 5). At low concentrations of chlorodinitrobenzene less mutagenicity was observed than at the same doses without S9. At higher doses a disproportionate increase in mutagenicity occurred. We have previously shown that the S9 fraction contains considerable amounts of GSH as well as GSH *S*-transferase activities (Summer et al., 1980). Addition of GSH to the test system completely abolished mutagenicity of chlorodinitrobenzene. These data indicate a close correlation between intracellular glutathione levels and chlorodinitrobenzene genotoxicity. Until mutagenicity of this chemical has been observed, it has been used topically to treat alopecia areata in man (Happle and Echternacht, 1977).

These observations are particularly important in view of the fact that chloroform also depletes glutathione and that chloroform toxicity can be antagonized by administration of SH-containing compounds such as cysteine which is a precursor of glutathione (Docks and Krishna, 1976). Reitz et al. (1978) who doubt the relevance of the high-dosed chloroform carcinogenicity tests in rats and mice reported evidence of a sudden change in the shape of the chloroform-response curve below 200 mg/kg. Before finally condemning chloroform, they ask for a complete study to evaluate carcinogenicity at low doses.

The following experiment further demonstrates the role of metabolic inactivation in toxicity of a chemical.

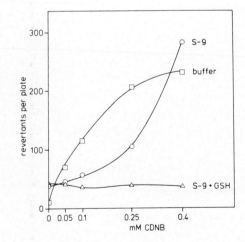

Fig. 5. Modification of chlorodinitrobenzene (CDNB) mutagenicity in the plate incorporation assay by glutathione conjugation. *S. typhimurium* were incubated with CDNB either in the presence of buffer or with postmitochondrial supernatant (S9) or S9 and 2.5 mM glutathione.

(2) Naphthalene: Protein binding and conjugation

Isolated hepatocytes metabolized naphthalene to water-soluble compounds (Schwarz et al., 1980). Metabolism of the hydrocarbon was linear for 1 h reaching a plateau after 2 h. During biotransformation of naphthalene reactive intermediates were formed which became irreversibly bound to cellular protein. Binding almost paralleled the increase in the formation of metabolites. Formation of water-soluble compounds and binding was due to metabolism since frozen-thawed cells in the presence of 1 mM SKF 525-A showed neither formation of water-soluble compounds nor binding to cellular macromolecules.

Conjugation with UDP-glucuronic acid and sulfate is one of the main pathways of naphthalene metabolism in hepatocytes (Bock et al., 1976). We inhibited these conjugation mechanisms by interfering with the synthesis of their respective co-factors. Addition of D-galactosamine reduces levels of uridinediphosphoglucuronic acid by trapping uridine-triphosphate and by inhibiting UDPG dehydrogenase activity (Bauer and Reutter, 1973). Sulfation is inhibited by incubation of the cells in sulfate-free medium which decreases synthesis of 3' phosphoadenosine-5-phosphosulfate (PAPS) and, thereby, sulfation (Schwarz, 1980). Incubation of hepatocytes in a sulfate-free medium in the presence of 3 mM D-galactosamine did not affect formation of water-soluble metabolites from naphthalene. However, a drastic increase in covalently bound metabolites was demonstrated (Fig. 6).

It is evident from this that cellular formation of reactive and toxic species of chemicals is frequently counteracted by metabolic detoxification. A disproportionate increase in the toxic effects can occur when high concentrations of a chemical are present which overcome inactivation processes.

In toxicology, such threshold levels on the basis of overwhelmed inactivation have long been established. For example, furosamide, a diuretic, is excreted predominantly unaltered in the urine when low therapeutic doses are given. Administration of high doses, however, overwhelm renal clearance and cause a disproportionate increase in toxic metabolites which react with macromolecules. Moreover, bromobenzene is metabolized in the liver to the reactive 3,4-bromobenzene oxide. This molecule is detoxified by glutathione con-

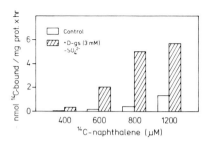

Fig. 6. Effect of 3 mM D-galactosamine (D-gs) and sulfate depletion on irreversible binding of [^{14}C] naphthalene. Hepatocytes were prepared from 200 g Sprague—Dawley rats by in situ perfusion with collagenase. Hepatocytes (1.25 × 10^6 cells) were incubated in Waymouth MB 752:1 medium in spinner flasks. Metabolism was stopped with ice-cold n-hexane. "Water soluble metabolites" of naphthalene were quantified by liquid-scintillation counting of the water phase plus neutralized NaOH extract of the organic phase containing the phenolic products. To determine irreversible binding of ^{14}C radioactivity, 0.5 ml of the cell suspension was sonified twice with 0.5 ml ethanol. The subsequent solvent extractions were performed as previously described (Hesse et al., 1978).

jugation. When high doses of bromobenzene are given, glutathione is depleted and the reactive metabolites interact with cellular macromolecules (Gillette, 1974a, b). Other chemicals for which there is evidence that high doses cause a disproportionate increase in toxicity and possibly carcinogenicity are aspirine, salicylamide, acetaminophen, styrene, ethylene glycol and aniline (Gehring et al., 1976).

(b) Impaired metabolic activation

On the other hand, potential carcinogens not only overwhelm inactivating capabilities but also inhibit metabolic activation reactions. The animal bioassay results on vinylchloride of Maltoni (1977), Lee et al. (1978) and Viola et al. (1971) suggest that inhalation beyond 150 ppm results in a smaller increase in carcinogenic response in rodents than that predicted from a linear extrapolation of the responses in the dose range of 0—150 ppm. Recent pharmacokinetic studies indicate that at high doses of vinyl chloride the mechanisms which metabolically activate this chemical are saturated. Bolt et al. (1980) observed no further increase of DNA and protein-bound vinyl chloride metabolites in rats beyond a certain level of exposure. A similar effect is reported with carbon tetrachloride by Ugazio et al. (1972). A preceding low dose of CCl_4 significantly increased the LD_{50} of rats when a high dose was given thereafter.

Modification due to species differences in metabolism

Moreover, species differences in the metabolic activation and inactivation systems further complicate extrapolation from high doses given in animal studies to the low doses to which man is exposed.

Several well-established examples indicate qualitative and quantitative species differences in the activating and inactivating metabolism interfering with the capability of the reactive intermediates to bind to the genetic material.

(a) Species differences in metabolic activation

(1) 2-Acetylaminofluorene

For example, 2-acetylaminofluorene (2-AAF), a strong hepatocarcinogen in male rats, is without effect in guinea pigs. 2-AAF is an indirect carcinogen. It requires aromatic N-hydroxylation forming the N-hydroxy-2-AAF (Miller et al., 1961). Since guinea pigs do not have this enzyme this species is insensitive to the carcinogen although they develop tumors when the N-hydroxy product is given (Miller et al., 1964).

Other species differences in the carcinogenicity of 2-AAF cannot be related to this activating mechanism. However, carcinogenicity of 2-AAF is also dependent on the formation of a sulfo-derivative. This reaction is catalysed by a N-sulfotransferase which differs quantitatively in several animal species. DeBaun et al. (1970a, b) determined the activity of this enzyme and were able to explain further the species differences of 2-AFF carcinogenicity (Table 1). Male rats, which are highly sensitive to this carcinogen, possess a high sulfotransferase activity. It is low in female rats and in mice of both sexes which are

TABLE 1

HEPATOCARCINOGENICITY OF 2-ACETYLAMINOFLUORENE AS RELATED TO SULFO-TRANSFERASE ACTIVITIES IN SEVERAL SPECIES

De Baun et al., 1970a, b.

		Hepatocarcinogenicity	Sulfo-transferase activity[a]
Rat	M	+++	23.0
	F	+	4.0
Mouse	M	+	0.5
	F	+	0.5
Guinea pig	M	—	0.3
Rabbit	M	—	2.0

[a] μmoles/0.01 ml supernatant per 20 min.

less susceptible than the male rat. The other species tested which have moderate or lower sulfotransferase activities are not known to produce hepatocarcinoma due to 2-AAF.

(2) Chloroform

Moreover, not only qualitative but also quantitative differences in the metabolizing rate exist between the different animal species and man. Several investigators have studied the metabolism of chloroform in rats and mice and have concluded that mice metabolize chloroform more rapidly than do rats (Pohl, 1979). The most complete studies are those of Brown et al. (1974) who administered a 60 mg/kg dose of chloroform orally to rats, squirrel monkeys and 3 strains of mice (Table 2). Very little unmetabolized chloroform was recovered from the mice. In the rat, 3 times as much unmetabolized chloroform was found, while in the monkey 13 times as much unaltered chloroform was exhaled. Fry et al. (1972) administered chloroform to human volunteers and found that 17—66% of the material was exhaled unchanged. However, the dose of chloroform used in these investigations was only 7 mg/kg. Thus, the relative amount of chloroform that would be metabolized at the higher doses used in the animal experiments may even have been overestimated. Based on these investigations, Reitz et al. (1978) suggested man to be the least sensitive of the species to the carcinogenic action of chloroform. This is also based on the data of Weiss et al. (1977) who observed that man normally metabolizes

TABLE 2[a]

INTERSPECIES DIFFERENCES IN CHLOROFORM METABOLISM AS DETERMINED BY URINARY EXCRETION OF THE UNCHANGED CHEMICAL

	Oral dose (mg/kg)	Unchanged $CHCl_3$ excreted (% of dose)
Mouse	60	6
Rat	60	20
Monkey	60	78
Man	7	17—66

[a] Modified from Reitz et al. (1978).

materials much more slowly than the small laboratory animals such as mice or rats. Consequently, a direct extrapolation from carcinogenicity tests on chloroform in mice or rats would overestimate the risk of man to this potential carcinogen.

(3) Halogenated ethylenes

In addition to chloroform, there are several other examples where the relative carcinogenicity of chemicals correlates well with their rate of activation to a reactive species. 1,1-Dichloroethylene is metabolized more extensively in the mouse than in the rat (McKenna et al., 1977). Correspondingly, tumors have been observed in mice exposed to 1,1-dichloroethylene (Maltoni, 1977), whereas in rats, although conflicting results appeared, no tumor formation seems to be related to this chemical (Fishbein, 1979; Maltoni, 1977; Rampey et al., 1977; Maltoni et al., 1977).

Trichloroethylene is carcinogenic in B6C3F1 mice but not in Osborn—Mendel rats (Natl. Cancer Inst., 1976). This again correlated with the greater capacity of mouse microsomes to catalyze binding of trichloroethylene metabolites to DNA (Bannerjee and van Duuren, 1978).

These data indicate that the status of metabolic activation systems in the different species significantly affects susceptibility to carcinogens in the animal and explain species differences.

(b) Species differences in metabolic inactivation

(1) Aflatoxin B_1

In addition, there is also evidence that species differences of inactivating mechanisms such as binding of the reactive intermediates with glutathione modify adverse effects of chemicals. The hepatotoxic and carcinogenic effects of aflatoxin B_1 are attributed to the reactions of metabolically formed AF B_1-2,3-oxide (Croy et al., 1978). The marked species differences in susceptibility to carcinogenicity could not be related so far to the different ability to generate the epoxide. Recently, Degen and Neumann (1978) found that AF B_1-epoxide is inactivated by conjugation with glutathione. Degen (1979) has since reported that mouse-liver preparations most effectively catalysed the formation of the glutathione conjugate. These results support the view that the lower susceptibility of mice is not a result of less effective activation but rather of more effective inactivation of the reactive AF B_1 intermediate due to the formation of glutathione conjugates.

(2) Styrene epoxide hydrolase

Oesch (1980), Oesch et al. (1973, 1974) and Jerina et al. (1977), studying the effects of microsomal epoxide hydrolase activity, further stressed the importance of considering species differences in metabolic inactivation. The hydrolase generally converts reactive epoxides into more stable dihydrodiols (Jerina et al., 1977) which are frequently further conjugated with glucuronic acid (Oesch, 1980). For example hydrolase activity is involved in the biotransformation of styrene (Leibman, 1975). It is highly suggested that hydrolase activity being present in the S9 fraction prevents styrene from being mutagenic in the plate-incorporation assay (Greim et al., 1977) although styrene oxide, the metabolite which is formed during metabolic activation of styrene, is a potent mutagen in this test (Mily and Garro, 1976).

Epoxide hydrolase generally plays an important role in the inactivation of many aromatic and olefinic compounds of industrial interest. Since man has a much higher hydrolase activity than most of the investigated laboratory animals (Oesch et al., 1974) it is to be expected that man is the least sensitive to such compounds.

We can conclude the following from the experimental data presented: Metabolism plays an important role in the carcinogenic or mutagenic activity of chemicals. It is species-dependent, can be influenced by the dose of the chemical administered, and thus has to be considered in extrapolation from the high doses given to the experimental animal to the low doses to which man is exposed.

Regarding metabolism, roughly 4 groups of chemicals can be differentiated:

(1) Compounds which undergo neither metabolic activation nor significant inactivation, e.g. methyl methanesulfonate.

(2) Compounds which only undergo metabolic activation, e.g. dimethyl- or diethyl-nitrosamine.

(3) Compounds which only undergo metabolic inactivation, e.g. N-methyl-N'-nitro-N-nitrosoguanidine.

(4) Compounds which undergo both metabolic activation and inactivation, e.g. vinyl chloride or chloroform.

For risk estimation, it is apparent that linear extrapolation from high to low doses and from one animal species to another is only possible for compounds of Group One which require neither metabolic activation nor inactivation.

Apparent species differences in the metabolic activation and inactivation system have to be considered concerning compounds of Groups Two to Four. Extrapolation can be further complicated when high doses of the chemical inhibit or overcome metabolic activation or inactivation.

Carcinogens acting by secondary mechanisms

Another group of compounds induces tumors by secondary mechanisms, for example, by suppressing production of hormones or by inducing necrosis at high doses which finally results in tumor formation.

Several chemicals inhibit function of the thyroid gland. These goitrogens induce tumors by a secondary mechanism. Thiourea blocks thyroxin synthesis by specific inhibition of iodine peroxydation (Davidson et al., 1979). As a consequence, insufficient amounts of thyroxin are produced by the thyroid gland leading to an increased hormonal stimulation of the thyroid by the hypophysis. This results in a hypertrophy of the thyroid, and after continuous exposure, thyroid adenoma appears (IARC, 1974). Finally, carcinomas of the thyroid are observed. At the lower doses of thiourea when no hypertrophy of the thyroid is observable, carcinomas of this organ have not been detected.

Carcinomas of the thyroid have been detected in rats and mice receiving a diet of 0.25% thiourea (Purves and Griesbach, 1947). This dose is equivalent to approx. 100 mg/kg daily. When administered as a drug, repeated daily intake of approx. 10 mg/kg was goitrogenic. Presently, thiourea is used as an industrial chemical only and sufficient measures to protect man at his working place have to be provided (Fishbein, 1979). However, workers would not be endangered by thiourea exposure unless exposed to concentrations affecting thyroid functions which can easily be detected.

Nitrilotriacetic acid and chlorothalonil, both excreted almost quantitatively via the kidney, induce urinary-tract tumors which have been preceded by tubular necrosis (Natl. Cancer Inst., 1977; Anderson and Kanerva, 1978; WHO, 1975; FAO, 1978; Hicks et al., 1975). Due to the almost 200-fold concentration of the glomerular filtrate in the tubular system, high intratubular concentrations of both compounds occur. When high doses are given to an animal, intratubular supersaturation of the chemicals with the consequences of precipitation and cristalluria appear. This frequently is associated with hematuria. Both symptoms, cristalluria and hematuria, have been observed when high concentrations of such chemicals are given. During long-term feeding of high doses to animals, tubular necrosis preceded renal-tumor formation (Anderson and Kanerva, 1978). At lower doses which do not induce tubular necrosis, neither cristalluria nor hematuria, renal necrosis or kidney tumors have been observed.

One might also refer to the tumor-inducing capability of saccharin and cyclamate. Both compounds have been shown to induce renal tumors at high doses (Hicks et al., 1975; Oser et al., 1975). Both substances, when given in large quantities to animals, increase the urinary pH substantially. This gives rise, at least in the case of saccharin as well as nitrilotriacetic acid, to subepithelial microcalculi (Anderson and Kanerva, 1978; Armstrong, 1977).

It has been demonstrated with other carcinogens that alkalization of the urine acts in a cocarcinogenic fashion, thus increasing tumor incidence. Therefore, the tumor inducing effect of such compounds may be a physicochemical one which is related to their alkalizing properties. Now if this is true, then this is a dose-related phenomenon. The doses humans would take or would be exposed to in the case of other similarly acting chemicals, would be virtually insignificant in that the buffer systems in the organism and in the urine would maintain the physiological pH of the urine. Again, extremely high doses given to animals are irrelevant to human exposure in such cases.

This sort of consideration requires more basic research than is presently available and, if true, we would be dealing with at least 2 classes of carcinogens. One group includes the classical type of the electrophilic reactants which react with target nucleophiles resulting in mutagenic or carcinogenic effects. For these mechanisms so far there is no clear evidence for a threshold. Another group of compounds which indirectly induce tumors by other mechanisms associated with extremely high doses are likely to have a threshold dose. These considerations have been taken into account only marginally so far. If this hypothesis can be clearly established, then it would be possible to deal with different classes of chemicals that induce tumors and one can start to deal with questions of acceptable dose levels. This would also imply that high doses of such compounds used in animal studies would be irrelevant to human exposure.

Effect of contaminants at high doses

Finally, the problem of the presence of impurities in chemicals being subjected to mutagenicity or carcinogenicity testing is pointed out as exemplified by trichloroethylene. This chemical has been reported to produce high incidence of hepatocellular carcinoma in both male and female mice but not in rats after high daily oral doses. (Memorandum, HEW, 1975). Pure trichloroethylene has been found to be slightly mutagenic in a modified bacterial liquid incubation system (Greim et al., 1975) and noncarcinogenic in rats, mice and hamsters (Henschler et al., 1980). Technical-grade trichloroethylene, however, contains several impurities. Using gas-chromatographic mass-spectroscopic analysis, Henschler et al. (1977) recently found 10 different compounds in trichloroethylene which amounted to 0.65% of the sample. Two of these, namely epichlorohydrin and epoxybutane, constitute strong alkylating agents which have been shown to be mutagenic (Henschler et al., 1977; Fishbein, 1976) and carcinogenic (IARC, 1976; van Duuren and Banerjee, 1976). It has been concluded that the carcinogenic effect of technical trichloroethylene is most probably due to these epoxides which are added to several brands for stabilization.

This is supported by the observations that hepatocellular carcinomas were produced in mice but not in rats (IARC, 1976). Mice show a relatively lower activity of epoxide hydrolase, the enzyme which detoxifies epoxides such as epichlorohydrin and epoxybutane (Oesch, 1973). In the carcinogenicity studies, the animals were exposed to high doses of trichloroethylene with concomitant high dose of the contaminants. Thus, when only high doses of a chemical become mutagenic or carcinogenic, possible contamination by impurities has to be considered as well.

Conclusions

First

Interspecies extrapolation and high to low doses extrapolation of chemicals undergoing metabolic activation and inactivation is most complicated. One has to consider: (1) Species differences in metabolic activation; (2) Species differences in metabolic inactivation; (3) Sufficient inactivating capacity may almost completely prevent interaction of reactive intermediates with the genetic material; (4) High amounts of a chemical may overwhelm inactivating processes with the consequence of disproportionately increased interaction with genetic material.

Second

Several chemicals act by secondary mechanisms which only become effective at doses much higher than those to which man is exposed by the chemical.

Third

When only high doses result in toxic effects including mutagenicity or carcinogenicity, contribution of impurities have to be considered as well.

Acknowledgement

The excellent secretarial help of Ms. Judy Byers is gratefully acknowledged.

References

Anderson, R.L., and R.L. Kanerva (1978) Hypercalciuria and cristalluria during ingestion of dietary nitrilotriacetate, Fd. Cosmet. Toxicol., 16, 569—574.
Andrae, U., and H. Greim (1979) Induction of DNA repair replication by hydroxyurea in human lymphoblastoid cells mediated by liver microsomes and NADPH, Biochem. Biophys. Res. Commun., 87, 50—58.
Armstrong (1977) Question and answer session, in: H.H. Hiatt, J.D. Watson and J.A. Winsten (Eds.), Origins of Human Cancer, Cold Spring Harbor Lab., pp. 1701—1708.
Banerjee, S., and B.L. van Duuren (1978) Covalent binding of the carcinogen trichloroethylene to hepatic microsomal proteins and to exogenous DNA in vitro, Cancer Res., 38, 776—780.
Bartsch, H., C. Malaveille, R. Montesano and L. Tomatis (1975) Tissue-mediated mutagenicity of vinylidine chloride and 2-chlorobutadiene in Salmonella typhimurium, Nature (London), 255, 641—643.
Bartsch, H., C. Malaveille, A. Barbin and G. Planche (1979) Mutagenic and alkylating metabolites of haloethylenes, chlorobutadienes and dichlorobutenes produced by rodent or human liver tissues, Evidence for oxirane formation by P-450 linked microsomal monooxygenase, Arch Toxicol., 41, 249—277.
Bauer, C., and W. Reutter (1973) Inhibition of diphosphoglucose dehydrogenase by galactosamine-1-phosphate and UDP-galactosamine, Biochim. Biophys. Acta, 293, 11—14.
Bock, K.W., G. van Ackeren, F. Lorch and F.W. Birke (1976) Metabolism of naphthalene to naphthalene dihydrodiol glucuronide in isolated hepatocytes and in liver microsomes, Biochem. Pharmacol. 25, 2351—2356.
Bolt, H.M., J.G. Filser, R.J. Laib and H. Ottenwälder (1980) Binding kinetics of vinylchloride and vinyl bromide at very low doses, Arch. Toxicol., Suppl. 3, 129—142.
Brown, D.M., P.F. Langley, D. Smith and D.C. Taylor (1974) Metabolism of chloroform — The metabolism of [^{14}C] chloroform by different species, Xenobiotica, 4, 151—163.
Carter, L.J. (1979) How to assess cancer risks, Science, 204, 811—816.
Commoner, B., A.J. Vithayathil, P. Dolara, S. Nair, P. Madyastha and G.C. Cuca (1978) Formation of mutagens in beef and beef extract during cooking, Science, 201, 913—916.
Cornfield, J. (1977) Carcinogenic risk assessment, Science, 198, 693—699.
Croy, R.G., J.M. Essigmann, V.N. Reinhold and G.N. Wogan (1978) Identification of the principal aflatoxin B$_1$—DNA adduct formed in vivo in rat liver, Proc. Natl. Acad. Sci. (U.S.A.), 75, 1748—1749.
Crump, K.S., D.G. Hoel, C.H. Langley and R. Peto (1976) Fundamental carcinogenic processes and their implications for low dose risk assessment, Cancer Res., 36, 2973—2979.
Davidson, B., M. Soodak, H.V. Strout, J.T. Neary, C. Nakamura and F. Maloof (1979) Thiourea and cyanamide as inhibitors of thyroid peroxidase: The role of iodide, Endocrinology, 104, 919—924.
DeBaun, J.R., E.C. Miller and J.A. Miller (1970a) N-Hydroxy-2-acetyl-aminofluorene sulfotransferase: Its probable role in carcinogenesis and in protein (methion-S-yl) binding in rat liver, Cancer Res., 30, 577—595.
DeBaun, J.R., E.Y.R. Smith, E.C. Miller and J.A. Miller (1970b) Reactivity in vivo of the carcinogen N-hydroxy-2-acetaminofluorene: Increase by sulfate ion, Science, 167, 184—186.
Degen, G.H. (1979) Metabolic inactivation of aflatoxin B$_1$-epoxide as a cause for the difference in aflatoxin-susceptibility of rat and mouse, An in vitro investigation, Naunyn-Schmiedeberg's Arch. Pharmacol., 307, Suppl. R15.
Degen, G.H., and H.G. Neumann (1978) The major metabolites of aflatoxin B$_1$ in the rat is a glutathione conjugate, Chem.-Biol. Interact., 25, 239—255.
Docks, E.L., and G. Krishna (1976) The role of glutathione in chloroform-induced hepatotoxicity, Exptl. Mol. Pathol., 24, 13—22.
FAO Plant Production Paper, 10 Suppl. (1978) Pesticides residues in food: 1977 evaluations, pp. 87—99.
Fishbein, L. (1976) Industrial mutagens and potential mutagens, I. Halogenated aliphatic derivatives, Mutation Res., 32, 267—307.
Fishbein, L. (1979) Potential Industrial Carcinogens and Mutagens, Elsevier, Amsterdam, pp. 211—217.
Fishman, J., D.K. Fukushima, J.O. O'Connor and H.T. Lynch (1979) Low urinary estrogen glucuronides in women at risk for familial breast cancer, Science, 204, 1089—1091.
Fry, B.J., T. Taylor and D.E. Hathway (1972) Pulmonary elimination of chloroform and its metabolite in man, Arch. Int. Pharmacodyn., 196, 98—111.
Gehring, P.J., G.E. Blau and P.G. Watanabe (1976) Pharmacokinetic studies in evaluation of the toxicological and environmental hazard of chemicals, in: Advances in Modern Toxicology — New Concepts in Safety Evaluation, Hemisphere, Washington.
Gillette, J.R. (1974a) A perspective on the role of chemically reactive metabolites of foreign compounds

in toxicity, I. Correlation of changes in covalent binding of reactive metabolites with changes in the incidence and severity of toxicity, Biochem. Pharmacol., 23, 2785—2794.

Gillette, J.R. (1974b) A perspective on the role of chemically reactive metabolites of foreign compounds in toxicity, II. Alterations in the kinetics of covalent binding, Biochem. Pharmacol., 23, 2927—2938.

Greim, H., G. Bonse, Z. Radwan, D. Reichert and D. Henschler (1975) Mutagenicity in vitro and potential carcinogenicity of chlorinated ethylenes as a function of metabolic oxirane formation, Biochem. Pharmacol., 24, 2013—2017.

Greim, H., D. Bimboes, G. Egert, W. Göggelmann and M. Krämer (1977) Mutagenicity and chromosomal aberrations as an analytical tool for in vitro detection of mammalian enzyme-mediated formation of reactive intermediates, Arch. Toxicol., 39, 159—169.

Haley, T.J. (1978) Chloroprene (2-chloro-1,3-butadiene) — What is the evidence for its carcinogenicity? Clin. Toxicol., 13, 153—170.

Happle, R., and K. Echternacht (1977) Induction of hair growth in alopecia areata with dinitrochlorobenzene, Lancet, II, 1002—1003.

Henschler, D., E. Eder, T. Neudecker and M. Metzler (1977) Carcinogenicity of trichloroethylene: Fact or artifact? Arch. Toxicol., 37, 233—236.

Henschler, D., W. Romen, H.M. Elsässer, D. Reichert, E. Eder and Z. Radwan (1980) Carcinogenicity study of trichloroethylene by longterm inhalation in three animal species, Arch. Toxicol., 43, 237—248.

Hesse, S., M. Mezger and T. Wolff (1978) Activation of [^{14}C]-chlorobiphenyls to protein-binding metabolites by rat liver microsomes, Chem.-Biol. Interact., 20, 355—365.

Hicks, R.M., J.St.J. Wakefield and J. Chowaniec (1975) Evaluation of a new model to detect bladder carcinogens or cocarcinogens: Results obtained with saccharin, cyclamate and cyclophosphamide, Chem.-Biol. Interact., 11, 225—233.

Hidebrandt, A.G., and I. Roots (1975) Reduced nicotinadenine dinucleotide phosphate (NADPH)-dependent formation and breakdown of hydrogen peroxide during mixed function oxidation reactions in liver microsomes, Arch. Biochem. Biophys., 171, 385—397.

IACR (1974) Monographs on the Evaluation of the Carcinogenic Risk of Chemicals to Humans, Vol. 6, Sex hormones, Lyon.

IARC (1974) Monographs on the Evaluation of Carcinogenic Risk of Chemicals to Man, Vol. 7, Some antithyroid and related substances, nitrofurans and industrial chemicals, Lyon, pp. 95—105.

IARC (1976) Monograph series on evaluation of carcinogenic risks of chemicals to man, Vol. 11, Lyon, pp. 131—139.

Interagency Regulatory Liaison Group (1979) Scientific bases for identification of potential carcinogens and estimations of risks, J. Natl. Cancer Inst., 63, 241—268.

Jaeger, R.J., R.B. Conolly and S.D. Murphy (1974) Effect of 18 h fast and glutathione depletion on 1,1-dichloroethylene-induced hepatotoxicity and lethality in rats, Exptl. Mol. Pathol., 20, 187—198.

Jerina, D.M., R. Lehr, M. Schaefer-Ridder, H. Yagi, J.M. Karle, D.R. Thakker, A.W. Wood, A.Y.H. Lu, D. Ryan, S. West, W. Levin and A.H. Conney (1977) Bay region epoxides of dihydrodiols: A concept explaining the mutagenic and carcinogenic activity of benzo[a]pyrene and benzo[a]anthracene, in: H.H. Hiatt, J.D. Watson and J.A. Winsten (Eds.), Origins of Human Cancer, Cold Spring Harbor Lab., pp. 639—658.

Jones, D.P., H. Thor, B. Andersson and S. Orrenius (1978) Detoxification reactions in isolated hepatocytes, J. Biol. Chem., 253, 6031—6037.

Lee, C.C., J.C. Bandhari, M. Winston, W.B. House, R.L. Dixon and J.S. Woods (1978) Carcinogenicity of vinyl chloride and vinylidene chloride, J. Toxicol. Environ. Health, 4, 15—30.

Leibman, K.C. (1975) Metabolism and toxicity of styrene, Environ. Health Perspect., 11, 115—119.

Lijinsky, W. (1977) Standard-setting for nitrites and nitrosamines, in: H.H. Hiatt, J.D. Watson and J.A. Winsten (Eds.), Origins of Human Cancer, Cold Spring Harbor Lab., pp. 1807—1811.

Maltoni, C. (1977) Vinyl chloride carcinogenicity: An experimental model for carcinogenesis studies, in: H.H. Hiatt, J.D. Watson and J.A. Winsten (Eds.) Origins of Human Cancer, Cold Spring Harbor Lab., Cold Spring Harbor, pp. 119—146.

Maltoni, C., G. Cotti, L. Morsi and P. Chieco (1977) Carcinogenicity bioassays of vinylidene chloride, research plans and early results, Med. Lavoro, 68, 241—262.

Maugh, T.H. (1978) Chemical carcinogenesis: How dangerous are low doses? Science, 202, 37—41.

McKenna, M.S., P.G. Watanabe and P.J. Gehring (1977) Pharmacokinetics of vinylidene chloride in the rat, Environ. Health Perspect., 21, 99—105.

Memorandum, Dept. of Health, Education and Welfare, Washington, (20.3.1975).

Miller, J.A., C.S. Wyatt, E.C. Miller and H.A. Hartmann (1961) The N-hydroxylation of 4-acetaminobiphenyl by the rat and dog and the strong carcinogenicity of N-hydroxy-4-acetaminobiphenyl in the rat, Cancer Res., 21, 1465—1473.

Miller, E.C., J.A. Miller and M. Enomoto (1964) The comparative carcinogenicities of 2-acetylaminofluorene and its N-hydroxy metabolite in mice, hamsters, and guinea pigs, Cancer Res., 24, 2018—2026.

Mily, P., and A.J. Garro (1976) Mutagenic activity of styrene oxide (1,2-epoxy ethyl benzene), a presumed styrene metabolite, Mutation Res., 40, 15—18.
Nagao, M., M. Honda, Y. Seino, T. Yahagi, T. Kawachi and T. Sugimura (1977) Mutagenicities of protein pyrolysates, Mutation Res., 53, 351—353.
National Cancer Institute (1976) Report on the Carcinogenesis Bioassay of Trichloroethylene, Bethesda, MD.
National Cancer Institute (1977) Bioassays of nitrilotriacetic acid (NTA) and nitrilotriacetic acid-trisodium salt-monohydrate (Na_3NTAH_2O) for possible carcinogenicity, NCI Tech. Rep. Ser. No. 6, January, DHEW Publication No. (NIH) 77—806.
Oesch, F. (1973) Mammalian epoxide hydrases: inducible enzymes catalysing the inactivation of carcinogenic and cytotoxic metabolites derived from aromatic and olefinic compounds, Xenobiotica, 3, 305—340.
Oesch, F. (1980) Species differences in activating and inactivating enzymes related to in vitro mutagenicity mediated by tissue preparations from these species, Arch. Toxicol., Suppl. 3, 179—194.
Oesch, F., N. Morris, J.W. Daly, J.E. Gielen and D.W. Nebert (1973) Genetic expression of the induction of epoxide hydrase and aryl hydrocarbon hydroxylase activities in the mouse by phenobarbital or 3-methylcholanthrene, Mol. Pharmacol., 9, 692—696.
Oesch, F., H. Toenen and H. Fahrländer (1974) Epoxide hydratase in human liver biopsy specimens: Assay and properties, Biochem. Pharmacol., 23, 1307—1317.
Oser, B.L., S. Carson, G.E. Cox, E.E. Vogin and S.S. Sternberg (1975) Chronic toxicity study of cyclamate: saccharine (10:1) in rats, Toxicology, 4, 315—330.
Plugge, H., and R.J. Jaeger (1979) Acute inhalation toxicity of 2-chloro-1,3-butadiene (Chloroprene): Effects on liver and lung, Toxicol. Appl. Pharmacol., 50, 565—572.
Pohl, L. (1979) Biochemical Toxicology of Chloroform, in: Hodgson, Bend and Philpot (Eds.), Reviews in Biochemical Toxicology, Elsevier/North-Holland, New York, pp. 79—107.
Ponemarkov, V., and L. Tomatis (1980) Long-term testing of vinylidene chloride and chloroprene for carcinogenicity in rats, Oncology, 37, 136—141.
Purves, H.D., and W.E. Griesbach (1947) Studies on experimental goitre, VIII. Thyroid tumors in rats tested with thiourea, Br. J. Pathol., 28, 46—53.
Rall, D.P. (1960) Difficulties in extrapolating the results of toxicity studies in laboratory animals to man, Environ. Res., 2, 360—367.
Rampey, L.W., J.F. Quast, C.G. Humiston, M.F. Balmer and B.A. Schwetz (1977) Interim results of two-year toxicological studies in rats of vinylidene chloride incorporated in the drinking water or administered by repeated inhalation, Environ. Hlth. Perspect., 21, 33—43.
Reitz, R.H., P.J. Gehring and C.N. Park (1978) Carcinogenic risk estimation for chloroform: An alternative to EPS's procedures, Fd. Cosmet. Toxicol., 16, 511—514.
Scanlan, R.A., J.F. Barbour, J.M. Hotchkiss and L.M. Libbey (1980) *N*-Nitrosodimethylamine in beer, Fd. Cosmet. Toxicol., 18, 27—29.
Schwarz, L. (1980) Modulation in sulfation and glucuronidation of 1-naphthol in isolated rat liver cells, Arch. Toxicol., 44, 137—145.
Schwarz, L., M. Mezger and S. Hesse (1980) Effect of decreased glucuronidation and sulfation on covalent binding of naphthalene in isolated rat hepatocytes, Toxicology, 90, in press.
Scientific Committee, Food Safety Council (1978) Quantitative risk assessment, Fd. Cosmet. Toxicol., 16, Suppl. 2, 109—136.
Spiegelhalder, B., G. Eisenbrand and R. Preussmann (1979) Contamination of beer with trace quantities of *N*-nitroso-dimethylamine, Fd. Cosmet. Toxicol., 17, 29—32.
Stockinger, H.E. (1977) Toxicology and drinking water contaminants, J. Am. Water Works Assoc., 69, 399—402.
Sugimura, T. (1978) Let's be scientific about the problem of mutagens in human cooked food, Mutation Res., 55, 149—152.
Sugimura, T., M. Nagao, T. Kawachi, M. Honda, T. Yahagi, Y. Seino, S. Sato, N. Matsukara, T. Matsushima, A. Shirai, M. Sawamura and H. Matsumoto (1977) Mutagens—carcinogens in food, with special reference to high mutagenic pyrolytic products in broiled foods, in: H.H. Hiatt, J.D. Watson and J.A. Winsten (Eds.), Origins of Human Cancer, Cold Spring Harbor Lab., pp. 1561—1577.
Summer, K.H., and W. Göggelmann (1980) 1-Chloro-2,4-dinitrobenzene depletes glutathione in rat skin and is mutagenic in *Salmonella typhimurium*, Mutation Res., 77, 91—93.
Summer, K.H., W. Göggelmann and H. Greim (1980) Glutathione and glutathione S-transferases in the Salmonella mammalian-microsome mutagenicity test, Mutation Res., 70, 269—278.
Summer, K.H., and H. Greim (1980) Detoxification of chloroprene (2-chloro-1,3-butadiene) with glutathione in the rat, Biochem. Biophys. Res. Commun., 96, 566—573.
Tannenbaum, S.R. (1980) Ins and outs of nitrites, The Sciences, 20, 7—9.
Tomatis, L. (1979) The predictive value of rodent carcinogenicity tests in the evaluation of human risks, Annu. Rev. Pharmacol. Toxicol., 19, 511—530.

Ugazio, G., R.P. Koch and R.O. Recknagel (1972) Mechanism of protection against carbon tetrachloride administration, Exptl. Mol. Pathol., 16, 281—285.

Van Duuren, B.L., and S. Banerjee (1976) Covalent interaction of metabolites of the carcinogen trichloroethylene in rat hepatic microsomes, Cancer Res., 36, 2419—2422.

Viola, P.L., A. Bigotti and A. Caputo (1971) Oncogenic response of rat skin, lungs and bones to vinyl chloride, Cancer Res., 31, 516—522.

Weiss, M., W. Sziegoleit and W. Förster (1977) Dependence of pharmacokinetic parameters on the body weight, Int. J. Clin. Pharmacol., 15, 572—575.

WHO (1975) Pesticide Residue Series No. 3, 1974 Evaluation of some pesticide residues in food, pp. 101—148.

FUNDAMENTAL MICROBIOLOGICAL ASPECTS OF THE AMES TEST
PANEL DISCUSSION

Chairman: J.P. Seiler (Wädenswil)
Panel Members. W. Göggelmann (Neuherberg); A. Grafe (Mannheim); M.H.L. Green (Brighton; I.E. Mattern (Rijswijk); J. Vollmar (Mannheim)

CONTENTS

Introduction
   by J.P. Seiler (Wädenswil).................................... 151
Influence of composition and treatment of the growth media on the yield of mutant colonies
   by J.P. Seiler (Wädenswil).................................... 155
The effect of spontaneous mutation on the sensitivity of the Ames test
   by M.H.L. Green (Brighton).................................. 159
Test cell problems in the Ames test
   by A. Grafe (Mannheim)....................................... 167
Culture conditions and influence of the number of bacteria on the number of revertants
   by W. Göggelmann (Neuherberg).............................. 173
Statistical problems in the Ames test
   by J. Vollmar (Mannheim) 179
Basis of evaluation of an Ames test
   by I.E. Mattern (Rijswijk).................................... 187
Summary and discussion
   by J.P. Seiler (Wädenswil).................................... 191

INTRODUCTION

J.P. SEILER

Swiss Federal Research Station, CH-8820 Wädenswil (Switzerland)

Following the discussions at the 1976 EEMS meeting in Gernrode and at a 1977 workshop in Munich (Mattern and Greim, 1978) a Collaborative Study with the Ames test was undertaken by 20 European laboratories. 3 compounds — 2-aminoanthracene, benzo[a]pyrene and N-nitrosomorpholine — were distributed to them from one source, eliminating compound-related differences, and were to be tested using a generally agreed upon protocol. The results of this study could finally be presented at a second workshop in Wädenswil (Seiler et al., 1980).

To tell it in the words of Dr. Grafe, who presented the results there, it could be "concluded that for all three test substances results were obtained which deviate very much ... Lack of mutagenicity, very high values concerning the mutagenic potency of a test compound and negative effects at very high concentrations were observable ... Large deviations among and within the individual laboratories were present". These statements are exemplified for the case of TA1538/benzo[a]pyrene (Fig. 1a) and for the case of TA1535 and 2-aminoanthracene (Fig. 1b). The outcome of this study was the more remarkable, as the compounds were well known mutagens and were tested uncoded. Dr. Grafe however, had not been satisfied only with a descriptive presentation of these results and in experiments with the help of Dr. Mattern he had been looking for possible causes of these divergencies. These additional experiments gave some clues about important details in the Ames test, which he then brought to our attention in Wädenswil. It was felt then, that knowledge about these technical details was too important a thing to be left without further investigation and most participants of the Wädenswil workshop agreed to continue their participation in one or another form in this field.

The importance of inter- and intra-laboratory variation concerning the Ames test results has also been recognized elsewhere and several studies have been published which tried, however, more to standardize the responses than to look at basic causes of variability (Dunkel, 1979; Topham, 1978; Cheli et al., 1980). While it is certainly possible to standardize the testing procedure between a few laboratories to such an extent that the test results become more or less uniform, it is nonetheless very important to investigate the parameters leading to variability in order to distinguish between necessary and unnecessary standardization. In this Panel Discussion results of these investigations will be

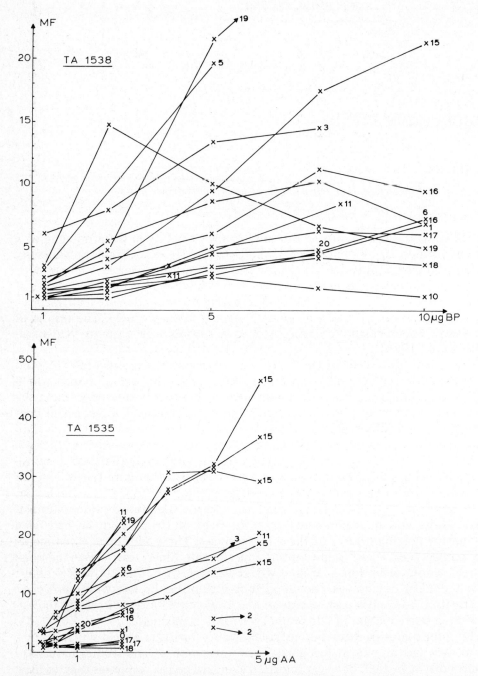

Fig. 1. Results from the Collaborative Ames Test Study in 19 European Laboratories (MF = mutation factor = induced mutants/spont. mutants). (a) Results with benzo[a]pyrene and strain TA1538 (30% S9). (b) Results with 2-aminoanthracene and strain TA1535 (30% S9).

presented and discussed. Deliberately left out of these investigations and also of this Panel Discussion is the whole area of metabolic activation by liver S9. The complexity of this mixture of dissolved and membrane-bound enzymes and cofactors precludes as yet a meaningful biochemical analysis. The often heard sentence: "The final proof of the S9 activity is the Ames test itself" is only the famous snake biting its tail when it comes to defining which factors are responsible for an optimal performance of the activation system and could or should be measured in order to assess the activating system. We will rather go back to the basics of the Ames test, which is a microbial test system and will discuss the microbiological fundamentals of this test system. We will thus go through the Ames test in the order: protocol and ingredients, "spontaneous" mutants, induction of mutants, statistical considerations, evaluation.

References

Cheli, C., D. DeFrancesco, L.A. Petrullo, E. McCory and H.S. Rosenkranz (1980) The Salmonella mutagenicity assay: Reproducibility, Mutation Res., 74, 145—150.

Dunkel, V.C. (1979) Collaborative studies on the Salmonella/microsome mutagenicity assay, J. Assoc. Off. Anal. Chem., 62, 874—882.

Mattern, I.E., and H. Greim (1978) Report of a workshop on bacterial in vitro mutagenicity test system, Mutation Res., 53, 369—378.

Seiler, J.P., I.E. Mattern, M.H.L. Green and D. Anderson (1980) Meeting report: Second European workshop on bacterial in vitro mutagenicity test systems, Mutation Res., 74, 71—75; Environ. Mutagen., 2, 97—100.

Topham, J.C. (1978) Interlaboratory variations of test results, Int. Symp. on Short-term Mutagenicity Test Systems for Detecting Carcinogens, Dortmund (FRG), Nov. 15—17.

INFLUENCE OF COMPOSITION AND TREATMENT OF THE GROWTH MEDIA ON THE YIELD OF MUTANT COLONIES

J.P. SEILER

Swiss Federal Research Station, CH-8820 Wädenswil (Switzerland)

Looking through a pile of papers using the Ames test as mutagenicity test system one will find invariably in the materials and methods section just a sentence like: "The Salmonella test was used as described by Ames", and reference is made to one of the methods papers of Ames' group. 3 Swiss laboratories working in one of the Ames test groups prepared step by step accounts of how the test is routinely done in the resp. laboratory. Differences occurred at nearly every level. An analogous observation has been made in a survey through a questionnaire sent out to all European laboratories performing the Ames test: of 83 laboratories answering the survey, 67 reported to be using the test with modifications (Würgler and Friederich, in preparation). Although these differences in laboratory practice would seem to be small and insignificant they may nevertheless influence the test result. Inasmuch as such differences include only different suppliers of the magnesium sulphate or the sodium chloride they may not contribute to differences in the results. However, the different suppliers could supply products which may even contain mutagens, as seems to be the case with several batches of nutrient broth (Ames et al., 1975, 1980). Such a mutagenic contaminant would add linearly to the number of colonies in the control as well as in the treated series; in the case of a weak mutagen to be tested — which is the case we are most concerned of here — this addition could well reduce the result to statistical non-significance.

Even worse results would be obtained, if the plastic plates had been sterilized by ethylene oxide, which has been reported by Nagao (as cited by Ames et al., 1975) and on one occasion also observed in our laboratory.

The influence of the histidine content in the agar plates on the number of colonies obtained would be considered as obvious. However, since Ames et al. (1975) give expected numbers of spontaneous colonies per plate, which, in many cases, are not obtained (see Fig. 1), it could be tempting, at least to the inexperienced novice in the field, to tamper with the histidine content of the top agar in order to obtain the published value of spontaneous mutants. There might be certain problems connected with variations in the histidine concentration as shown in Fig. 2. A higher concentration of histidine allows the bacterial growth on the plate for more generations than a lower one; on the one hand the number of spontaneous mutants per plate will increase, but the

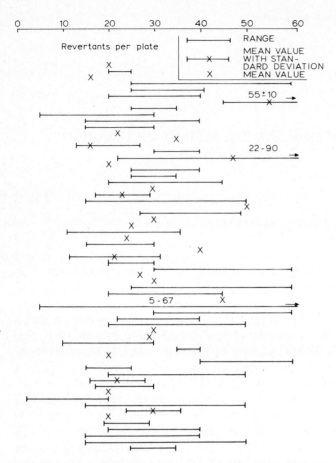

Fig. 1. Number of spontaneous revertants per plate in strain TA98 reported by a number of European laboratories (from Würgler and Friederich, in preparation).

number of induced revertants could also be influenced by the chemical nature of the test substance. A stable compound like sodium azide (see Fig. 2a) will remain active throughout the whole growth phase of the bacteria and will thus under all conditions of altering histidine concentrations increase the number of colonies by a more or less constant factor. A very unstable compound, which might exert its influence only during a limited period of the bacterial growth on the plate, would on the other hand theoretically be expected to become seemingly less mutagenic with increasing histidine concentration. As a third example, a substance which needs to be activated by the bacterial metabolism, may appear to increase in mutagenic potency when higher histidine concentrations are used (see Fig. 2b).

Another possibility of influencing the test result by the technical handling lies in the sterilization procedure. If the agar is autoclaved together with the salts and the glucose, the number of "spontaneous" mutants increases — pointing to the production of mutagenic substances — , and at the same time the number of induced mutants decrease — pointing to the occurrence of toxic substances in the agar (see Fig. 3). This kind of reaction may happen when the agar solution after the combination of the (separately autoclaved) ingredients is left at elevated temperatures for extended periods of time before pouring the plates. Such effects thus tend to flatten the concentration—effect curve and consequently to obscure a weak mutagenic activity.

Wrong conditions for the sterilization of the glucose by autoclaving on the other hand may lead to the production of co-mutagenic substances (Okamoto and Yoshida, 1980), and the choice of nature and amount of the solvent for

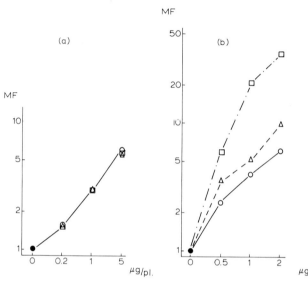

Fig. 2. Influence of the histidine concentration on the mutagenic response of (a) sodium azide in TA100; (b) 2-nitrofluorene in TA98. MF = Mutation factor = number of induced mutants per plate/number of spontaneous mutants per plate. 0.05 µg his/plate; 0.1 µg his/plate = normal amount; 0.2 µg his/plate.

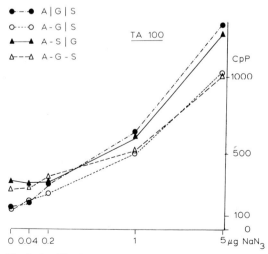

Fig. 3. Sterilization procedure of the minimal agar and its influence on the mutagenic response. A, agar; G, glucose; S, salts; |, autoclaved separately, —, autoclaved together.

mutagenic compounds is still another area where the yield of induced mutants can be greatly influenced (Anderson and McGregor, 1980).

In summary, it can be stated, that no 2 laboratories perform the Ames test in a completely identical manner; however, even seemingly minor deviations can have an effect on the outcome of the test.

References

Ames, B.N., J. McCann and E. Yamasaki (1975) Methods for detecting carcinogens and mutagens with the Salmonella/mammalian-microsome test, Mutation Res., 31, 347—364.

Ames, B.N. (1980) Supplement to the methods paper, personal communication to recipients of strains.

Anderson, D., and D.B. McGregor (1980) The effect of solvents upon the yield of revertants in the Salmonella/activation mutagenicity assay, Carcinogenesis, 1, 363—366.

Okamoto, H., and D. Yoshida (1980) Enhancement of mutagenicity of acetylaminofluorene by heating product of glucose in alkaline solution, Agric. Biol. Chem., 44, 439—441.

THE EFFECT OF SPONTANEOUS MUTATION ON THE SENSITIVITY OF THE AMES TEST

M.H.L. GREEN

MRC Cell Mutation Unit, University of Sussex, Falmer, Brighton BN1 9QH (Great Britain)

Pre-existing and plate mutants

When we detect induced mutation in the Ames test (Ames et al., 1975) we do so against a background of spontaneous mutation. I will try to describe briefly what we would expect to be the effects of spontaneous mutation on the sensitivity of the test. Fig. 1 shows the last few generations of growth of an Ames strain in an overnight culture.

If a mutation occurs during the last generation it will give rise to one mutant cell. A mutation in the previous generation will be half as likely since there will be half as many cells. On the other hand it will give rise to 2 mutant cells. The generation before, there will be one quarter the chance of mutation occurring but such a mutation will give rise to 4 mutant cells, and so on for each generation. Therefore if a mutation occurs in an overnight culture, it may give rise to one or many colonies when the culture is used in an Ames test, depending on the generation in which the mutation occurred. On the other hand if a mutation occurs during growth in the agar overlay of an Ames test it will give rise to

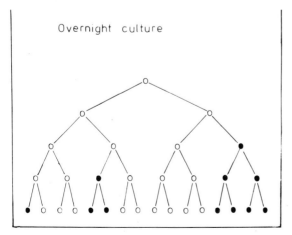

Fig. 1. Mutations arising during the last few generations of growth of an overnight culture. It can be seen that one event can give rise to one or many mutant cells. ○, non-mutant cell; ●, mutant cell.

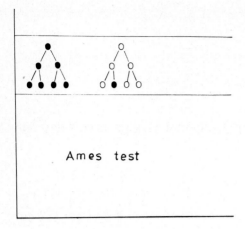

Fig. 2. Mutations arising during the last few generations of growth in the top agar of an Ames plate. It can be seen that one mutation will give rise to one colony. ○, non-mutant cell; ●, mutant cell.

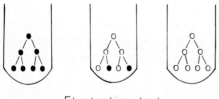

Fig. 3. Mutations arising during the last few generations of growth in a fluctuation test. It can be seen that one mutation will give rise to one turbid tube, but that if an additional mutation arises in the same tube, it will not be detected. ○, non-mutant cell; ●, mutant cell.

one colony only (Fig. 2). Similarly in a fluctuation test (Green et al., 1976) one mutation gives rise to one turbid tube (Fig. 3). It should be noted that in the fluctuation test, if more than one mutation occurs in a tube, only one event will be scored. This leads to a theoretical loss of sensitivity in comparison to the Ames test. In practice, the loss of sensitivity is slight, but a more important problem is that it is not possible to obtain adequate dose—response curves with the fluctuation test since above a very low dose, all the tubes are turbid.

If mutations arise with normal sampling error but one mutation, depending on when it arises, may give rise to one or many mutant cells, the result is that the number of His^+ mutant bacteria in an overnight culture is likely to be very variable. This variability was the basis of the proof by Luria and Delbruck (1943) that bacteria mutate and if it were not found, it would be evidence that His^+ reversion in Salmonella was not a genuine mutation. The number of pre-existing bacteria in an overnight culture can readily be determined by plating in the absence of histidine. Obviously, in an Ames test, the number of pre-existing mutants present will be proportional to the number of cells plated. It is therefore sometimes thought that the problem of pre-existing mutants can be overcome by simply using fewer cells, but it is fairly easy to see that this is not a rational suggestion.

The majority of the spontaneous mutants that we see in an Ames test, however, are not pre-existing mutants but result from mutations that have occurred on the plate while the bacteria grow in the presence of the small histidine supplement. Since mutations should arise at random and one mutation should give rise to one colony, the number of colonies corresponding to plate mutations should be rather reproducible and subject only to normal sampling error. Furthermore, mutations are most likely during the last generations of growth when most cells are present. For this reason the number of mutants arising on the plate will be strongly dependent on the final number of cells on the plate and largely independent of the number of cells plated (Demerec and Cahn, 1953). The final number of cells on the plate depend, of course, on the histidine supplement.

It is for these considerations, that in the collaborative study on genetic drift in Ames strains, we have proposed comparing spontaneous mutability in 10-fold diluted cultures. The variable contribution of pre-existing mutants should be greatly reduced, whereas the reproducible contribution of plate mutants should be largely unaffected.

Measurement of induced mutation

Clearly, the ideal situation for measuring induced mutation is one in which we detect spontaneous and induced mutations arising over the same time period and each spontaneous or induced event gives rise to one and only one colony. Almost the worst situation is in the "Treat and Plate" test (Green and Muriel, 1976) where one induced mutation gives one colony, but a spontaneous mutation in the overnight culture can give rise to one or many colonies depending on when it arises. In addition we measure induced mutations arising over a short period against a background of spontaneous mutations arising over a considerably longer period. The treat and plate method is useful and appropriate for certain types of experiment, but it is not the method of choice for detecting weak mutagenic effects.

Before discussing how closely the Ames and fluctuation test approach the ideal, several complications must be mentioned. Firstly a period of growth is required during which DNA damage leads to an alteration in the genetic code. This process is known as mutation fixation. A further period of growth is then required before an alteration in the genetic code leads to production of an altered gene product and an altered phenotype. This process is known as mutation expression. Estimates of the average length of time required for mutation fixation vary. In excision-proficient strains of *E. coli* it is short (0—0.5 generations). In excision-deficient WP2 strains of *E. coli* average times for mutation fixation after UV between 0.5 and 1.5 generations have been reported, with a shorter time in the presence of casamino acids (Bridges and Munson, 1968; Doubleday et al. 1975). An estimate of the time required for mutation expression in *E. coli* is 0.5—1 generation (Bridges and Munson, 1968). I am not aware of any data on mutation fixation or mutation expression in Ames strains, although one might expect times to be similar to those in excision-deficient *E. coli*.

It is reasonably easy to measure the time required for induced mutation fixation and expression, by giving a short mutagenic treatment and determining how long it takes for the extra mutants to be expressed. Obviously this cannot be done for spontaneous mutation, and it would be very difficult (not impossible) to measure expression time. In the absence of data, it is reasonable to assume that the time for mutation expression will be similar for spontaneous and induced mutation (0.5—1 generation). This means that when we measure spontaneous mutation in bacteria, the value we obtain is an underestimate, since the mutation we see arose in a smaller population of cells than the one we measure.

There is no data whatever on the time required for spontaneous mutation fixation in bacteria. If we make the plausible assumption that spontaneous mutations arise primarily through the inaccuracy of DNA replication, we would expect that spontaneous mutation would occur almost entirely in growing cells but that fixation of spontaneous mutations would be immediate. The small supplement of histidine present in the Ames or fluctuation tests allows growth of the bacteria, and mutation fixation and expression will occur during this growth. The practical effect is merely that mutations arising in the last generation or two of growth of the bacteria will not be seen. If, as seems likely, induced mutation fixation is slower than spontaneous mutation fixation, spontaneous mutation will be measured in a larger population of cells, and the test will be correspondingly less sensitive.

In our tests, we make the important assumptions that the accuracy of DNA replication, and the length of time required for spontaneous mutation expression are constant and will not be affected by the types of agent screened for mutagenicity. Normally these assumptions will be entirely reasonable, but when screening such things as food or water samples, they may be distinctly risky. Both accuracy of DNA replication and mutation expression time might well be influenced by the physiological state of the bacteria, and a treatment that affected the general physiology of the bacteria might readily influence the yield of mutants.

A further and more serious complication applies to the Ames test. Because the mixture is applied as a thin agar overlay many of the constituents can diffuse into the bottom agar and their concentration will decrease during the time of study. Such things as the level of histidine and of test agent will decline. Some constituents of the S9-activating system may diffuse away, thus altering the pattern of metabolism. Diffusion of metabolites of the test agent may also be important (Forster et al., 1980). The problem of diffusion should not, of course, arise with the fluctuation test. Normally it will not matter particularly in the Ames test, but it does mean that the number of cells plated and their physiological state may exert a greater influence on results than might otherwise be expected.

Interpretation of results

The simplest criterion for declaring an Ames test positive is a doubling (or 2.5 × or 3 × increase) in the number of mutants per plate in the treated series over the number of mutants per plate in the control series. Dr. Mattern will explain how difficult it is to relate on this basis results between strains with different spontaneous mutation rates and Dr. Vollmar will explain how the criterion is not related to statistical significance. From what I have said above, it will be realised that a doubling in the number of mutants per plate is not even directly related to the ability of an agent to double the spontaneous mutation rate. Apart from the variable contribution of pre-existing spontaneous mutants to the test, it is extremely likely that spontaneous mutation will occur in an increasing cell population while the induced mutation rate will decline as the test agent diffuses out of the top agar layer, and the metabolising system loses efficiency. In order to double the number of mutants per plate, a test agent will have to very much more than double the mutation rate of an Ames strain. Exactly how much more will depend on the stability of the agent, its toxicity, the way it is metabolised and the rate of mutation fixation.

Normally Ames test positives are unambiguous, but when marginal results are being explained away, they may indeed be an artefact but it is worth remembering that a doubling in the number of mutants per plate in an Ames test, may correspond to very much more than a doubling of the actual mutation rate.

Improving test sensitivity

It is often argued that bacterial mutation assays are already too sensitive. This is really only another way of saying that scientists, industrialists, administrators and the general public are too stupid to be allowed to evaluate weak or marginal effects. It is not an argument I find convincing. Increasing test sensitivity helps to distinguish weak effects from statistical and methodological artefacts and enables us to work closer to relevant dose levels.

It also is often argued that mutagenicity testing should be carried out by standard protocols. This is an admirable sentiment provided that the standard protocol is appropriate for the agent under test. As I have pointed out above, however, a given increase in the number of mutants per plate in an Ames test in different experiments may mean entirely different things in terms of statistical significance or actual increase in mutation rate. If we wish to detect the smallest possible *bona fide* increase over the spontaneous mutation rate, we may well have to change the standard protocol. I give 3 examples.

(1) *Toxic, weakly mutagenic agents.* For such agents diffusion in the Ames test becomes a major disadvantage. If a non-toxic level of agent is added to the top agar overlay, it will diffuse away and the final concentration will be too low to be mutagenic. If a higher level of agent is used, it will kill the test bacteria. One solution is to use the fluctuation test, where diffusion is not a problem. A second solution is to add the test agent to the bottom agar layer of an Ames test, rather than the agar overlay. In this way, for instance, Bridges and Mottershead (1971) were able to show that 8-methoxypsoralen was mutagenic to the Ames strain TA1538 in the dark, although negative results had previoulsy been reported (Scott, 1976).

(2) *Highly unstable agents.* When a highly unstable agent is tested by a standard Ames protocol, it is effectively a "treat and plate test" in terms of sensitivity. It has the twin disadvantages that induced mutation arising over a short

period is measured against spontaneous mutation arising over a long period, and that induced mutations will give rise to one colony, whereas spontaneous mutations in the overnight culture may give rise to many colonies. It is generally thought that both these problems are insoluble, but this is not necessarily true. Clearly, if one is to treat an adequate population of cells, and allow for mutation expression, spontaneous mutation will occur over a longer period than induced mutation. But it is not necessary for one spontaneous mutation to give rise to more than one colony. In the fluctuation test an extra step can be inserted. Initially the tubes will be inoculated with a fairly small number of bacteria so that no pre-existing mutants are added. The bacteria are then allowed to grow up to exhaust a small supplement of histidine. Although spontaneous mutation can occur during this period, one mutation can not give rise to more than one turbid tube. The test agent is then added, together with additional histidine, to allow mutations induced to be expressed. The same effect might be obtained in an Ames test by pouring a second agar overlay, or by impinging the test bacteria on a membrane filter as described below.

(3) *Agents toxic to growing bacteria.* One of the most desirable features of the Ames test is that the bacteria being treated are actively growing, for many agents are more mutagenic to growing bacteria. An agent such as penicillin, however, is far more toxic to growing cells and a test by standard protocol is likely to be fairly pointless. Normally an insensitive treat and plate test is then used. If such an agent can be inactivated, a multistep fluctuation test, as described above, might be effective. Otherwise, I would suggest impinging the test bacteria on a membrane filter, rather than incorporating them in the agar overlay. If the filter is placed face up on an agar plate, the medium passes through the filter to the bacteria. The bacteria can form a lawn and mutant colonies can be scored. Since the bacteria are fixed in position on the filter, one mutation will give rise to one colony. The filter can be transferred from a plate containing a small supplement of histidine, to a plate of non-growth medium containing the test agent, and back to a histidine supplemented plate for mutation expression.

Conclusion

The standard Ames protocol (Ames et al., 1975) is excellent, and in the vast majority of cases is probably the simplest and most appropriate method of testing for mutagenicity. I am not against standard protocols as such, but neither am I in favour of using them when they are not appropriate. I hope that it can be seen that there is a rational basis to measuring mutation. Results fit theory, adequately, if not perfectly. It will sometimes be appropriate to modify the standard protocol and sometimes a test by the standard protocol will be almost meaningless. I hope that we can identify many of these situations and when they arise design an appropriate experiment, or recognise that an appropriate experiment has been performed.

Acknowledgement

I would like to thank Professor B.A. Bridges for his helpful suggestions.

References

Ames, B.N., J. McCann and E. Yamasaki (1975) Methods for detecting carcinogens and mutagens with the Salmonella/mammalian microsome mutagenicity test, Mutation Res., 31, 347—364.

Bridges, B.A., and R.P. Mottershead (1977) Frameshift mutagenesis by 8-methoxypsoralen in the dark, Mutation Res., 44, 305—312.

Bridges, B.A., and R.J. Munson (1968) Mutagenesis in *Escherichia coli*: evidence for the mechanism of base change mutation by ultraviolet radiation in a strain deficient in excision repair, Proc. R. Soc. (London) B, 171, 213—226.

Demerec, M., and E. Cahn (1953) Studies of mutability in nutritionally deficient strains of *Escherichia coli*, I. Genetic analysis of five auxotrophic strains, J. Bacteriol., 65, 27—36.

Doubleday, O.P., B.A. Bridges and M.H.L. Green (1975) Mutagenic DNA repair in *Escherichia coli*, II. Factors affecting loss of photoreversibility of UV-induced mutations, Mol. Gen. Genet., 140, 221—230.

Forster, R., M.H.L. Green and A. Priestley (1980) Optimal levels of S9 fraction in the Ames and fluctuation tests: apparent importance of metabolites from top agar, Carcinogenesis, 1, 337—346.

Green, M.H.L., and W.J. Muriel (1976) Mutagen testing using Trp^+ reversion in *Escherichia coli*, Mutation Res., 38, 3—32.

Green, M.H.L., W.J. Muriel and B.A. Bridges (1976) Use of a simplified fluctuation test to detect low levels of mutagens, Mutation Res., 38, 33—42.

Luria, S.E., and M. Delbruck (1943) Mutations of bacteria from virus sensitivity to virus resistance, Genetics, 28, 491—511.

Scott, B.R. (1976) Failure to detect a mutagenic activity of 8-methoxypsoralen (in the dark) in strains of Salmonella typhimurium and *Escherichia coli*, Mutation Res., 40, 167—168.

TEST CELL PROBLEMS IN THE AMES TEST

A. GRAFE

Boehringer Mannheim GmbH, Medizinische Forschung, Sandhofer Strasse 116, D-6800 Mannheim 31 (F.R.G.)

The results of the European Collaborative Ames Test Study, involving 20 laboratories, which were discussed at the 2nd Workshop in Wädenswil in 1979, showed great variations between the laboratories which in some cases were outside the tolerable range of standardized microbiological experiments.

Upon analysis of all the experimental results I concluded that one of the causes for the considerable variations was to be found in the test cell content used, since differing culture conditions lead to large differences in the bacteria counts.

Since the cell contents used in the tests were largely not recorded and in other cases the cell masses were determined photometrically, we have endeavoured in the meantime to gain a better understanding of the relationship between the test cells used and the test results.

My contribution to this discussion is based on the following: *Firstly*, in April 1979 we discussed the results of the European Collaborative Ames Test Study in Wädenswil. *Secondly*, according to the results of this discussion we formed a working group with regard to test cell problems, and *Thirdly*, we did experiments in my laboratory concerning the microbiology of the test strains.

For our problem it was first necessary to standardise our entire testing procedure as strictly as possible in order to make the results between the 5 laboratories coincide. In this way it was possible that out of 42 individual tests only the results of 1 test was outside the range which is tolerated in microbiological collaborative studies; this agreement is shown in Fig. 1 for TA1535 and sodium azide.

On the basis of this strict standardisation with the same test strain, the same culture conditions with the same flasks and shaking frequencies, identical methods and the same material, the test cell content was varied in 4 steps in tests with TA100 and TA1535 using the substance N-methyl-N'-nitro-N-nitrosoguanidine. The results with the 2 strains were identical.

In the tests with undiluted material more than 10^8 bacteria per plate were used and at the dilution of 1:125 the content of bacteria per plate was about 1×10^6. In order to obtain an exact determination of the pre-existing revertants, it was necessary to remove the histidine originating from the nutrient

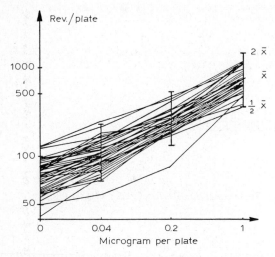

Fig. 1. Standardised tests with sodiumazide and TA1535 in 5 laboratories.

broth culture, which contained 1.8%. To this purpose the bacteria suspensions were washed. Fig. 2 shows the results with strain TA100.

Subdivision of the test findings corresponding to the test cell dilutions can be seen. The highest numbers of induced revertants were obtained with the highest numbers of test cells; the lowest numbers of induced revertants are found associated with the lowest numbers of test cells. The variations with the various cell contents overlap one another in such a way that a variation range in the cell content in the ratio 1:5 corresponds approximately to a doubling of the variation range of the induced revertants per plate. Furthermore, a threshold can be seen below which the testing of a substance is uncertain because a concentration-dependent increase in revertant induction does not occur. This threshold is placed within the section of the results in which the test cell numbers per plate were lower than 1×10^8 but higher than 1×10^7.

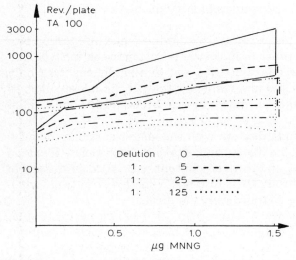

Fig. 2. The areas of results with different test cell numbers in 5 laboratories.

A more exact determination of this threshold was done graphically (see Fig. 3). For this purpose the arithmetic means of all test results were calculated from: the test cell content of each individual dilution step, the induced revertants from the same dilution step and the same substance concentration, the spontaneous revertants and the pre-existing revertants.

It can be seen that the substance was tested with most certainty when at least 4×10^7 test bacteria per plate were used. The lower the number of test cells, the more findings lay below the doubling threshold of the spontaneous mutants.

From this figure the most important points are: The higher the number of test cells the higher the number of substance-induced revertants. There is no parallelism between the number of cells per plate on the one side and the number of induced and spontaneous (plate + pre-existing) revertants on the other side. There is no parallelism between the number of substance-induced revertants and any type of non-substance induced revertants. It seems to exist a parallelism between the number of test cells and the number of pre-existing revertants.

These facts have several consequences: (1) Optimal test results can only be obtained with the use of maximal test cell numbers. (2) If the evaluation is based on the concentration—effect relationship and the requirement, that the number of substance-induced revertants must exceed twice the spontaneous value, then the substance can be tested with most certainty when at least 4×10^7 test bacteria per plate were used. (3) A mutagenic substance can also be discovered using test cell numbers below the maximum attainable if only the concentration—effect relationship of induced revertants is used as a basis for the assessment. (4) The number of spontaneous revertants is more important as a criterion for the test strain than as an exact basis for the statistical evaluation of Ames test results. Therefore, the calculation of mutation factors on the basis of the spontaneous revertants is not useful. (5) Deviations in the test cell content are reflected in the results. (6) In collaborative Ames test studies that are not strictly standardised it cannot be expected that the results will lie within the close range regarded as normal for standardised microbiological collaborative studies.

Fig. 3 shows even more problems, however. The non-parallelism of the courses and the non-linearity of the induced revertant counts do not permit a quantitative evaluation of the test results, and they necessitate critical consideration of the test strains from microbiological aspects.

If one searches the Ames test literature for its microbiological basis, one is without success. Although one finds the requirement — repeated from one publication to another — that the test should be carried out with more than 10^8 bacteria per plate and the cells should be taken from an overnight culture, no substantiation can be found.

"Overnight-culture" is firstly not a scientific definition and secondly not a suitable replacement for a clear statement regarding which growth phase the test cells should be taken for the test. Out of 4 different phases (see Fig. 4) in a statical culture of bacteria only a limited section of the growth curve is suitable for the Ames test. From the point of view that high numbers of test cells are necessary, bacteria from the end of the log-phase up to the end of the stationary phase are suitable.

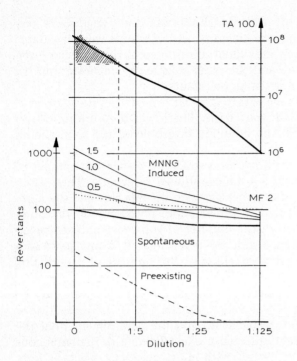

Fig. 3. Test cell numbers in standardized Ames tests.

But a second aspect has to be considered. In the Ames test, bacteria from a first culture in nutrient broth medium are brought into a second on minimum agar. The behaviour of the bacteria in the second culture is dependent upon the time at which cells were taken from the first culture, i.e. upon the decision as to what growth phase would be worked with. For a single substance, this decision might be unimportant, but for routine tests it may have consequences. As there is a relationship between the growth phase of the first culture and the duration of the lag-phase in the second culture, it is indicated to use bacteria from the phase with the highest physiological activity and the shortest lag-phase on minimal plates. Only in this way we will have the least trouble with the test strains and the stability of test substances.

Fig. 4 shows, that the requirement for high cell number is met when the bacteria are removed during the course of growth between the end of the loga-

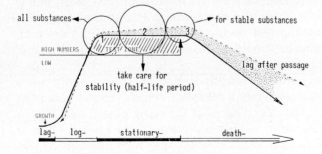

Fig. 4. Growth phases and Ames test.

rithmic and the beginning of the decline phase. For bacterio-physiological reasons, however, the bacteria should be taken at a time as close as possible to the logarithmic phase, i.e. at the beginning of the stationary phase. If there are special reasons for proceeding differently in the Ames test, then the reasons should also be known and appropriate test requirements be made.

Fig. 5 shows the growth courses determined by us for the standard Ames test strains using a shaking water bath. It can be seen that under our culture conditions all 5 test strains enter the stationary phase between 6 and 8 h and that at this point of time the highest numbers of test cells can be obtained. Therefore I cannot see any advantage to wait 16 or 18 h for taking cells — but there exist some disadvantages, of which I may mention only two.

In a 16–18 h culture a part of the test cells has already reached the end of the stationary or the beginning of the decline phase. In the physiological condition of the bacteria at this point only a part of the test cells might be able to overcome the lag-phase on minimum agar, although to this agar histidine and biotin are added, and most of the added cells might die off, leaving the experimenter in the unwarranted belief of having used a large number of cells.

I may direct special attention to strains TA98 and 1538. From a 16-h culture only a small part of the bacteria is able to grow on minimal plates. This experimental error may be the reason that only few substances show their mutagenicity in tests with these 2 strains.

The other disadvantage is connected with the stability of the test substance. Within a 2-h lag-phase many test substances may have lost much of their biological activity, leading also to false negative results and wrong conclusions.

From the aspect of the test cell problems, my conclusions are as follows: (1) The hitherto used Ames test recommendations disregard fundamental microbiological facts. (2) The test cell contents should be recorded by viable cell counts and not by the measurement of optical densities. (3) Negative Ames test results can only be evaluated if the work was carried out with at least 5×10^7 test bacteria per plate taken at the beginning of the stationary growth phase. (4) The concentration–effect relationship is the decisive criterion in the Ames test.

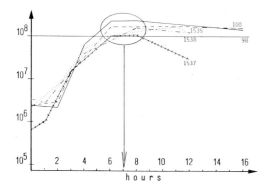

Fig. 5. Test cell growth in the shaking culture.

CULTURE CONDITIONS AND INFLUENCE OF THE NUMBER OF BACTERIA ON THE NUMBER OF REVERTANTS

W GÖGGELMANN

Abteilung für Toxikologie, Gesellschaft für Strahlen- und Umweltforschung, Ingolstädter Landstr. 1, D-8042 Neuherberg, Post Oberschleissheim (FRG)

In our first experiments performed in 1977 we used overnight cultures as described by Ames and measured their optical density and bacteria content. The number of cells plated differed for example for TA100 from $0.2-1.3 \times 10^8$ and for TA98 from $0.5-1.8 \times 10^8$ bacteria from experiment to experiment. For a routine test these bacteria titres varied considerably and furthermore did not always reach the titre of at least 10^8 bacteria per plate as recommended by Ames. With modified culture conditions all tester strains reached the stationary phase after 8 h with comparable bacteria titres. Only using bacteria cultures from the end of the logarithmic phase we found well reproducible numbers of bacteria and of spontaneous revertants per plate (Table 1). These data indicated that the number of bacteria plated is determined by the culture conditions used.

The problem of the number of bacteria per plate and their influence on the

TABLE 1

RANGES OF BACTERIA AND SPONTANEOUS REVERTANTS PER PLATE[a]

Strain		Bacteria per plate $\times 10^6$	Spontaneous revertants per plate
TA1535	1977	181–307	38–83
	1978	144–236	31–63
	1979	196–262	52–74
TA1538	1977	237–329	25–47
	1978	155–271	14–28
	1979	241–315	25–36
TA100	1977	166–252	113–156
	1978	146–196	108–137
	1979	162–220	110–134
TA98	1977	183–247	27–49
	1978	169–229	22–36
	1979	170–240	28–44

[a] Bacteria were cultured under modified growing conditions (4-baffled flasks and a shaking frequency of 90 per min) and used at the end of the logarithmic phase (6–8 h).

number of revertants was discussed during the 2nd Ames test workshop in Wädenswil in 1979. Therefore we compared our growing conditions (4-baffled flasks, shaking frequency of 90 per min) with growing conditions frequently used for the Ames test (flasks, shaking frequency of 60 per min). Under both conditions bacteria cultures were grown up for 16 h and every 2 h we determined the optical density of the culture, the number of bacteria on nutrient broth agar plates and the number of revertants.

Figs. 1 and 2 show that under both conditions characteristic growth curves are obtained. When bacteria are shaken slowly their number increases continuously and reaches the maximal cell number between 12 and 16 h (TA100 0.6×10^8 and TA98 1.2×10^8 bacteria plated). Under our culture conditions (rapid shaking) all tester strains reach the stationary phase within 8 h with an average content of 2.5×10^8 bacteria per plate. At longer times of growth the

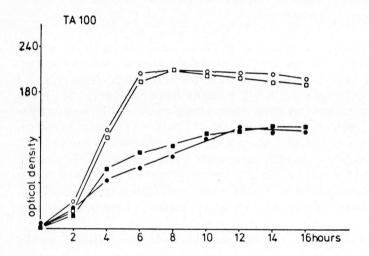

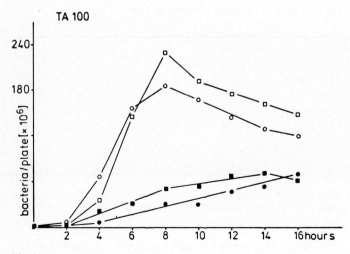

Fig. 1.

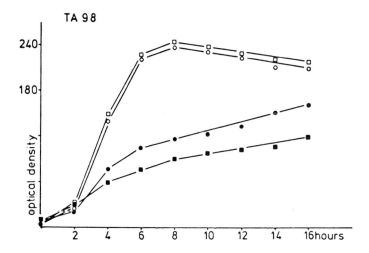

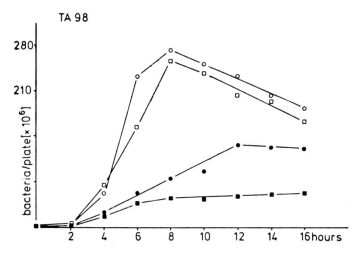

Fig. 2.

Figs. 1 and 2. Growth curves of TA100 and TA98 with different culture conditions. The optical density of the growing cultures was measured with a Klett–Summerson Photoelectric Colorimeter (filter 500–570 mµ). The number of bacteria was determined on nutrient broth agar plates (0.1 ml of the 10^{-4} to 10^{-6} dilutions). Each value is the mean value from 4 plates. Culture conditions: ○, □, 4-baffled flasks and a shaking frequency of 90 per min; ●, ■, flasks and a shaking frequency of 60 per min.

number of bacteria plated decreases but the optical density remains rather constant. This indicates that the total number of bacteria in the culture is constant but the number of viable (colony-forming) cells decreases. Therefore the determination of the optical density is not a reliable measure for the number of viable bacteria in a culture.

Figs. 1 and 2 show that the culture conditions do influence the number of cells used in the Ames test and that highest numbers of bacteria can be received only at the end of the logarithmic phase under well-defined growing conditions.

Furthermore we could demonstrate that not only the number of spontaneous revertants is influenced by the number of bacteria per plate but the number of induced revertants as well. This relationship was shown with 4 different mutagens on a poster contribution at the 9th EEMS meeting in 1979 and additionally was studied by 5 different laboratories in a collaborative study. These results have been already discussed by Dr. Grafe.

Moreover, in our laboratory we investigated if bacteria taken from different growth phases will influence the number of induced and spontaneous revertants in the Ames test. Figs. 3—5 show the results obtained with 2 primary (sodium azide, NaN_3, and 1-chloro-2,4-dinitrobenzene, CDNB) and 1 secondary (2-aminoanthracene, 2-AA) mutagens.

Using bacteria cultures of titres lower than 5×10^7 bacteria the bacterial growth on the Ames plates is inhibited with nearly all test substances. However

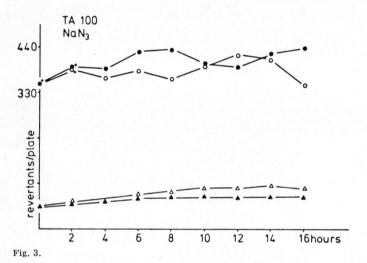

Fig. 3.

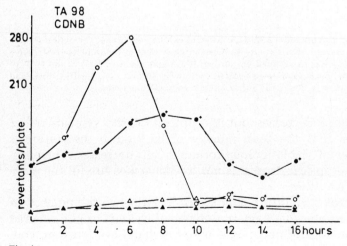

Fig. 4.

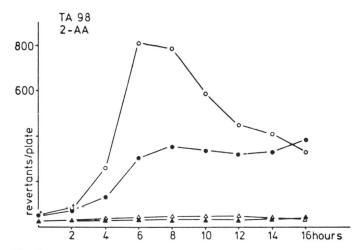

Fig. 5.

Figs. 3, 4 and 5. Relationship between the number of bacteria and growth phases and the number of spontaneous (△, ▲) and induced (○, ●) revertants, sodium azide, NaN_3; 1-chloro-2,4-dinitrobenzene, CDNB; 2-aminoanthracene, 2-AA. Each value is the mean value from 3 plates. Culture conditions as Fig. 1, + reduced bacterial growth on the Ames plates.

with CDNB the bacterial growth is reduced during the whole test period if bacteria are taken from a slowly shaken culture. In such a case it is absolutely necessary to perform the test with bacteria from the end of the logarithmic phase (rapidly shaken culture).

With sodium azide (Fig. 3) the number of induced revertants is nearly equal during the whole test period, independently of the mode of bacteria culturing.

Contrary, with bacteria of rapidly shaken cultures and CDNB or 2-AA as mutagens the number of induced revertants is dependent on the number of bacteria plated and on the growth phase. For example in the presence of CDNB (Fig. 4) bacteria of a 6-h culture (2.3×10^8 bacteria per plate) show the highest number of induced revertants. After 8 h of shaking the bacteria titre (2.7×10^8 bacteria per plate) is higher and the culture is at the beginning of the stationary phase but the number of induced revertants is decreased drastically. With bacteria cultures grown longer than 10 h the number of induced revertants approaches the spontaneous level. The viable bacteria content of a 12-h culture (2.3×10^8 bacteria per plate) is the same as of a 6-h culture.

A similar effect is shown in Fig. 5 with 2-AA as mutagen. A 6-h growing culture with a titre of 2.3×10^8 bacteria per plate produces 830 revertants, whereas a 12-h culture with the same titre produces only 410 revertants per plate.

From these results it is obvious that some mutagens show much higher effects in bacteria from the logarithmic phase, e.g. replicating cells, than in bacteria from the stationary or decline phase.

Based on these data we summarize: (1) The growing conditions influence the number of cells used in the Ames test. (2) The determination of the optical density is not a reliable measure for the number of viable bacteria in a culture. (3) Each laboratory should perform the Ames test in a standardized procedure under well-defined culture conditions. (4) The Ames test has to be done with more than 5×10^7 bacteria per plate. (5) The highest number of bacteria is received at the end of the logarithmic phase. (6) Depending on the mutagen the number of induced revertants can be influenced by the number and by the growth phase of bacteria used.

STATISTICAL PROBLEMS IN THE AMES TEST

J. VOLLMAR

Boehringer Mannheim GmbH, Medizinische Forschung, Sandhofer Strasse 116, D 6800 Mannheim 31 (FRG)

(1) Before starting with the statistical evaluation of the results of an experiment there is a typical pathway which must be followed (Fig. 1). The statistician can only help the experimenter in a satisfactory way if, and only if, the first step from the "biological problem" to the "statistical problem" is correct. But from the remarks of Drs. Green, Grafe and Göggelmann one can conclude that at the moment a quantitative solution of the biological problem "Ames Test" is not possible. I came to the same conclusion by the following statistical finding (Vollmar, 1979).

The number of revertants is not Poisson-distributed

In Fig. 2 we can see a log range — log mean plot of all results with 3 plates/concentration from the European Collaborative Ames-Test Study 1977/78. There are too many experiments having a variation too large for a Poisson distribution. In the case of a Poisson distribution 90% of all observations should lie between the 2 straight lines designated 0.05 and 0.95 (Pettigrew and Mohler, 1967). As, however, can be seen in Fig. 2, this does not hold for mean revertant numbers greater than 100. The larger variability cannot be

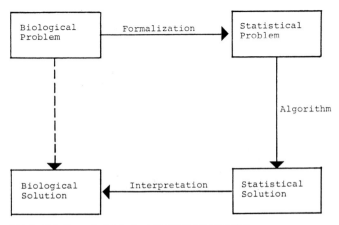

Fig. 1. Solution of a biological problem by statistics.

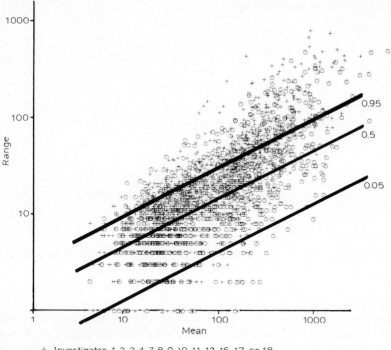

+ Investigator 1, 2, 3, 4, 7, 8, 9, 10, 11, 12, 16, 17 or 18
○ Investigator 5, 6, 13, 15 or 20

Fig. 2. Relationship between mean number of revertants and range of number of revertants (3 plates/dose).

explained by an interlaboratory variation. The intralaboratory variation, as can be seen in Fig. 3 (TA100) and Fig. 4 (TA1535), where only the observations of 1 laboratory are plotted, looks very similar to the interlaboratory variation in Fig. 2. Another possible explanation could be the different behavior of plate and pre-existing revertants. But as can be seen in Fig. 5 there are too many observations lying above the 0.95 line (29 instead of 9). Other possible reasons for deviation from the Poisson distribution may, for instance, be coagulation on the plate, errors during pipetting or during plating. On evaluating all the results of the collaborative Ames Test study, both an "over-dispersion" of the results (above "0.95"), and deviations which are too small are observed. The latter was usually the case when counting was not carefully performed (making the data "fit").

No common function is known which would fit all concentration—effect relationships possible in the Ames test

There may be one or more concentration ranges for which there is a linear concentration—effect relationship, but there are also ranges for which the concentration is non-linear. However, it is also noticeable that the curves for the various bacterial strains are of the same type but run at distinctly different levels.

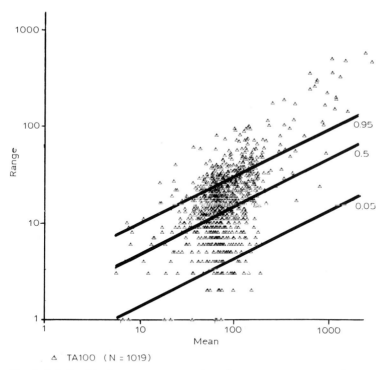

Fig. 3. Relationship between mean number of revertants and range of number of revertants (3 plates/concentration) in 1 laboratory (strain TA100).

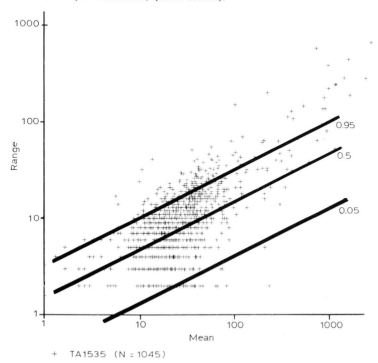

Fig. 4. Relationship between mean number of revertants and range of number of revertants (3 plates/concentration) in 1 laboratory (strain TA1535).

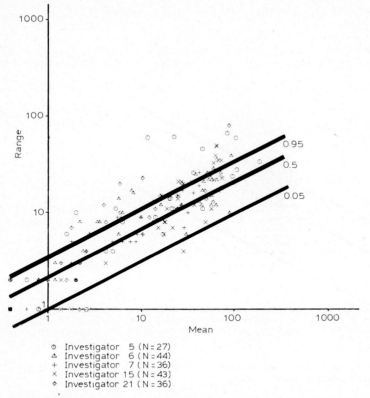

Fig. 5. Relationship between mean and range of number of pre-existing mutants (5 plates/concentration) in 5 laboratories.

Using the Ames test no quantitative statement can be made

One of the previously mentioned factors in the results of an Ames test is the total number of bacteria used (Grafe, 1979, 1980). In Table 1 can be seen that the number of spontaneous revertants seems to be practically independent of this number of bacteria/plate. Dr. Grafe has also shown us in his Fig. 3 that the number of spontaneous revertants seemed to be independent of the dilution, i.e. the number of test cells. But the number of pre-existing mutants correlates very closely with the number of bacteria used. From this finding it does not seem reasonable to suggest a factor (e.g. 2) which should be reached for a range of concentrations in order to classify the compound investigated as "positive in the Ames test", because the proportion of pre-existing mutants will strongly influence the observed factor.

Because of the different systematic interfering factors no quantitative statement can be expected of the Ames test. But qualitative statements are possible and these statements can be based on the results of statistical tests.

(2) We should apply distribution free (non-parametric) methods because at the moment we do not know the underlying distribution functions. The results have to be characterized by at least 2 suitable parameters, a parameter of location and a parameter of dispersion. As a measure of location the median

TABLE 1

RELATIONSHIP BETWEEN BACTERIAL CONTENT AND NUMBER OF REVERTANTS IN CONTROL PLATES (MEDIANS)

Bacteria/plate ($\times 10^7$)	TA100			TA1535		
	Number of revertants	Mutation frequency ($\times 10^7$)	N	Number of revertants	Mutation frequency ($\times 10^7$)	N
<1	99	5000	15	45	2127	18
1–4.9	68	192	10	24	53	5
5–9.9	69	97	180	38	54	130
≥10	67	49	65	33	26	115

value can be recommended. This value is not as dependent on outliers as is the mean value. The appropriate robust parameter of dispersion is the median deviation, which is defined as the median of the absolute differences between the single values and the median. (If a normal distribution could be assumed, the standard deviation is approx. 1.5 × median deviation.)

According to the different types of concentration—effect relationship it is necessary to choose different statistical tests in order to be able to make a decision as efficiently as possible. Fig. 6 shows 4 different types of concentration—effect curves. In case A it is possible to apply a "normal" linear regression analysis, but the heteroscedasticity of the number of revertants for the various concentrations has to be taken into account. This could be done by a suitable transformation of the individual data (for a Poisson distribution there would be a square-root transformation). In general, the 3 other types are observed, and not type A. A suitable method for all types is the Kruskal—Wallis test (Kruskal and Wallis, 1952). This test is a multi-sample generalisation of the two-sided Wilcoxon rank sum test (Wilcoxon, 1945). This method cannot be as efficient as a one-sided analogue which would apply in cases B and C. Suitable methods for these types of monotonic increasing concentration—effect curves are distribution-free tests for ordered alternatives. A multi-sample generalisation of the one-sided Mann—Whitney—Wilcoxon two-sample test (Mann and Whitney, 1947) is the Jonckheere test. In Table 2, the statistical results obtained with the Jonckheere and the Kruskal—Wallis tests on 713 Ames tests from the Collaborative Study are compared. The superiority of the Jonckheere test is clearly evident. In the case of benzo[a]pyrene the larger number of 30 positive Kruskal—Wallis tests compared with the Jonckheere test will be more extensively investigated. Fig. 7 shows these results graphically. Here you can clearly see the concentration—effect curves which first increase monotonically, and then decrease monotonically. If only the monotonically increasing part of the curve is considered, the Jonckheere test would be the most suitable statistical evaluation method.

One can therefore recommend the Jonckheere test as the routine statistical method for qualitative evaluations. For some situations, more refined statistical techniques could be applied. For Type C in Fig. 6 the test procedure of Chacko and Shorack (Chacko, 1963; Shorack, 1967) is more efficient, but the disadvantage of this method is that the critical values for small sample sizes have not yet been tabulated.

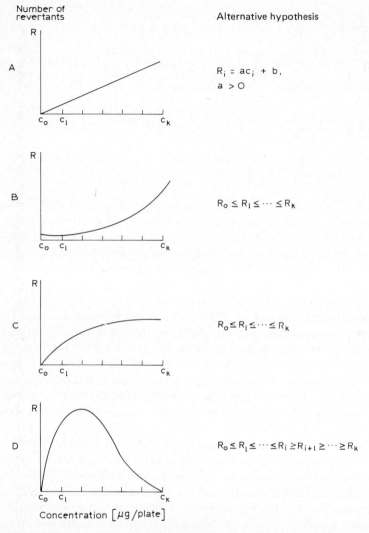

Fig. 6. Different types of concentration—effect curves for mutagenic compounds.

The choice of the statistical test must, however, be made before the results become known. If a significant effect is present, then it is necessary to decide for which minimal concentration the effect is significantly different from the 0-concentration. If there are k non-zero concentrations, at least k multiple comparisons (concentration versus control) must be carried out. In the literature (Hollander and Wolfe, 1973; Lehmann, 1975) distribution-free multiple comparisons based on Kruskal—Wallis rank sums are proposed for cases B, C and D. I see a different possibility by carrying out multiple comparisons with one-sided (for cases B and C) or two-sided (in case D) linear rank tests. A computer program for this statistical evaluation method was published (Stucky and Vollmar, 1976). This means that these methods could be applied without too many difficulties. It should, however, be borne in mind that a correction of the significance level must also be made (for instance according to Bonferroni).

(3) The main problem of planning of each experiment is to determine the sample size. This number depends on the probabilities of the errors of the first and of the second type, on the minimal effect which still may be regarded as being biologically relevant, on the choice of concentrations to be tested, and last, but not least, on the selected statistical method.

TABLE 2

COMPARISON OF THE STATISTICAL RESULTS OBTAINED WITH THE JONCKHEERE AND THE KRUSKAL—WALLIS TEST ($\alpha = 0.05$) FROM 713 AMES TESTS

		Result of Jonckheere test		
		not significant	significant	
Result of Kruskal—Wallis test	not significant significant	79 30	45 144	Benzo[a]pyrene ($n = 298$)
	not significant significant	39 5	40 170	2-amino-anthracene ($n = 254$)
	not significant significant	65 2	25 69	N-Nitrosomorpho-line ($n = 161$)

Significant, $p \leqslant 0.05$.
Not significant, $p > 0.05$.

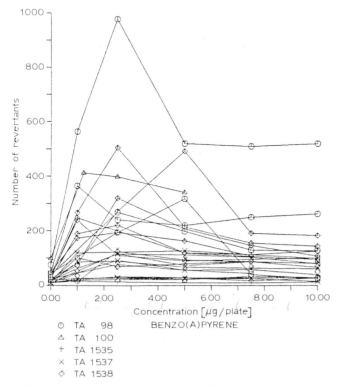

○ TA 98   BENZO(A)PYRENE
△ TA 100
+ TA 1535
× TA 1537
◇ TA 1538

Fig. 7. Concentration—effect curves (Jonckheere Test, not significant; Kruskal—Wallis Test, significant).

TABLE 3

MINIMAL NUMBER OF PLATES NECESSARY TO REACH p-VALUES $\leqslant 0.05$ FOR APPLYING THE MULTIPLE MANN—WHITNEY—WILCOXON TEST

	k				
	1	2	3	4	5
"one-sided"	3	4	4	5	5
"two-sided"	4	5	5	5	5

k, number of concentrations $\neq 0$.

At the moment I do not know any theoretically correct solution to the sample size problem. If, however, one plans to use the Mann—Whitney—Wilcoxon test for a pair-wise comparison, then the least number of plates necessary is shown in Table 3. These minimal numbers of plates result from the fact that when using fewer plates no significant results may be achieved for the given significance levels. At least 3—5 plates should be used per concentration, as is shown in Table 3.

(4) As a consequence derived from the foregoing aspects, the following is valid: *The Jonckheere test supplies a sufficiently qualitative evaluation of the Ames test when using 3—5 plates/concentration. Suitable parameters of location and dispersion are the median and median deviation.*

As long as no quantitative evaluation method is developed, results of studies on quantitative effects in this biological test system cannot be interpreted definitely. Especially, collaborative studies will not give a quantitative result because of the many — often unknown and confounded — factors influencing the Ames test.

References

Chacko, V.J. (1963) Testing homogenicity against ordered alternatives, Ann. Math. Statist., 34, 945—956.
Grafe, A. (1979) Results of the study without any special statistical evaluation and interpretation of the results from a microbiological point of view using additional experiments, in: A. Grafe and J. Vollmar (Eds.), The Collaborative Ames-Test Study 1977/78 — Analysis and Interpretation of the Results, Mannheim.
Grafe, A. (1980) These Proceedings, pp. 167—171.
Hollander, M., and D.A. Wolfe (1973) Non-Parametric Statistical Methods, Wiley, New York.
Jonckheere, A.R. (1954) A distribution-free k-sample test against ordered alternatives, Biometrika, 41, 133—145.
Kruskal, W.H., and W.A. Wallis (1952) Use of ranks in one-criterion variance analysis, J. Am. Statist. Assoc., 47, 583—621.
Lehmann, E.L. (1975) Non parametrics — statistical methods based on ranks, Holden-Day, San Francisco.
Mann, H.B., and D.R. Whitney (1947) On a test of whether one of two random variables is stochastically larger than the other, Ann. Math. Statist., 18, 50—60.
Pettigrew, H.M., and W.C. Mohler (1967) A rapid test for the Poisson distribution using the range, Biometrics, 23, 685—692.
Shorack, G.R. (1967) Testing against ordered alternatives in model I analysis of variance, normal theory and non-parametric, Ann. Math. Statist., 38, 1740—1752.
Stucky, W., and J. Vollmar (1976) Exact probabilities for tied linear rank tests, J. Statist. Comput. Simul., 5, 73—81.
Vollmar, J. (1979) Collaborative Ames-test study — Statistical aspects on the presentation and evaluation of the results of the joint study, in: A. Grafe and J. Vollmar (Eds.), The Collaborative Ames-Test Study 1977/78 — Analysis and Interpretation of the Results, Mannheim.
Wilcoxon, F. (1945) Individual comparisons by ranking methods, Biometrics, 1, 80—83.

BASIS OF EVALUATION OF AN AMES TEST

I.E. MATTERN

Medical-Biological Laboratory TNO, P.O. Box 15, 2280 AA Rijswijk (The Netherlands)

The first point to be considered in the evaluation of an Ames test is whether the test has been properly planned and conducted, taking into account the variables that can influence the test results; the other panel members have already made some remarks on this subject.

When the test has been performed well, the next question is how do we interpret the results of the test, i.e. the number of His$^+$ revertants/plate at various doses of test substance, in terms of mutagenic activity of the substance; a second question is how to compare the mutagenic activity of different substances. It is important that we know how to do this for the bacterial strains, before we can try to extrapolate from these results meaningful statements on the mutagenic or carcinogenic potency of the test substance for humans.

There are two points that I would like to bring to your attention for discussion:

(1) Should one use the number of induced His$^+$ revertants per plate (= total number minus spontaneous number) or the relative increase of revertants (= ratio between the (induced) number and the spontaneous number) as indicator of mutagenic activity?

(2) Should one routinely measure the survival of the cells in the Ames assay, and take this into account when expressing the mutagenic activity, e.g. as number of induced mutants/survivor/dose unit?

Ad 1. Both methods are commonly used; many people express the mutagenic potency of a test substance as number of induced His$^+$ revertants/nmole (taking the number from a linear portion of the concentration/response curve, if present); in deciding about "mutagenic or not" a factor of 2 increase over the spontaneous value is generally taken as the minimal criterium. Now, apart from the technical and statistical uncertainties, inherent in the test system (as has been pointed out by the other panel members), there is a more fundamental problem concerning the use of both methods, which becomes apparent when one wants to compare the mutagenic activity of a substance on strains with different spontaneous mutability, like TA98 and TA100. The same applies when one wants to compare 2 substances that induce only frameshift mutations or base-pair substitution mutations, respectively.

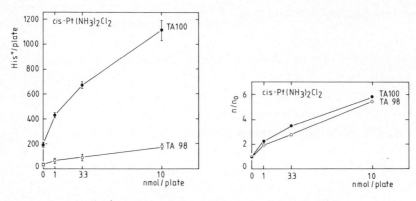

Fig. 1. Number of His⁺ revertants/plate at various doses of cis-Pt(NH$_3$)$_2$Cl$_2$.

Fig. 2. Relative increase of the number of His⁺ revertants/plate at various doses of cis-Pt(NH$_3$)$_2$Cl$_2$.

I would like to illustrate this point with an example from our own work. Fig. 1 shows the number of His⁺ revertants/plate (n) induced by cis-Pt(NH$_3$)$_2$Cl$_2$ in TA98 and in TA100; in Fig. 2 the same data are plotted, now relative to the number of spontaneous mutants (n/n_0). In Table 1 the corresponding data are given for 10 nmoles cis-Pt(NH$_3$)$_2$Cl$_2$/plate.

It is clear that a much higher response is obtained in TA100 than in TA98 when one compares the number of His⁺ revertants/plate; comparing the increase over the spontaneous number, however, approximately the same response is obtained in both strains.

This difference becomes important when we wish to interpret the results in terms of mechanisms. Is it allowed to conclude from the fact that cis-Pt(NH$_3$)$_2$Cl$_2$ at the same dose induces a much higher number of base-pair substitutions (in TA100) than frameshift mutations (in TA98), that this substance predominantly causes mutations of the former type? Or should one take the other view and say that cis-Pt(NH$_3$)$_2$Cl$_2$ is equally effective in the induction of the 2 types of mutations?

The answer depends on one's views on (and knowledge of) the mechanism of induction of spontaneous and induced revertants respectively, considering the type of damage done to DNA and the way the cell handles this damage (repair processes, "intrinsic" mutability of the strain etc.). In general, knowledge on these subjects is virtually absent, so a well founded choice cannot be made at this moment.

TABLE 1

His⁺ REVERTANTS INDUCED BY 10 nmoles cis-Pt(NH$_3$)$_2$Cl$_2$

His⁺ revertants/plate		TA98	TA100
Spontaneous	(n_0)	30	190
Total	(n)	180	1110
Induced	$(n - n_0)$	150	920
Ratio I	(n/n_0)	6.0	5.8
Ratio II	$(n - n_0)/n_0$	5.0	4.8

Ad. 2. Measurement of survival becomes of importance when the toxicity of the compounds plays a significant role. For instance, when one wants a more quantitative statement on the mutagenic activity of a compound, in many cases the toxicity should be taken into account. This is especially important for weak mutagens which have to be used in toxic doses and for the comparison of the mutagenic potency of mutagens differing in their ratio mutagenicity/toxicity.

We have to realise that the number of His$^+$ revertants which we observe on the plate is the end result of a series of complicated events which are in general ill-understood. In fact, the Ames test can be considered as a kind of black box and we do not know exactly what happens inside this box. Inside this box, the following processes may occur: (1) permeation into the cell → internal dose; (2) reaction to DNA/other cell constituents → amount of DNA damage; (3) repair processes, error-prone and error-free ⇄ mutation / lethality.

Although little is known about the precise nature of these processes, it can be argued that the outcome of the experiment, i.e. the observed number of His$^+$ revertants/plate, is determined by the final number of mutants produced and the relative survival of normal and mutant cells.

For example, a chemical that has a low toxicity (let's say 80% survival at a certain dose) induces a certain number of revertants at this dose; another chemical, that is equally mutagenic but more toxic (let's say 40% survival at the same dose) would give only half the number of induced revertants per plate.

Thus, in order to get a more accurate estimate of the mutagenic potency, the survival of the cells should be measured under the conditions and at the doses used in the Ames assay. Mutagenic potency could be expressed as the number of induced mutants/survivor/dose unit, or as the dose that gives a certain number of induced His$^+$ mutants/survivor.

This procedure might be relevant for studies on the relation between structure and function of chemicals and for the comparison of the mutagenic potency of different chemicals on a certain bacterial strain. When one only wants a "yes" or "no" answer, the number of induced His$^+$ revertants per plate can give a good approximation, provided doses are considered that give a sufficiently high survival (and keeping the uncertainties of the test system in mind).

To conclude, the following questions might be discussed: In order to express and compare the mutagenic activity of chemicals in various bacterial strains, should one (1) use the "absolute" number of His$^+$ revertants/plate or the relative increase over the spontaneous value? (2) determine the surviving fraction of the cells?

SUMMARY AND DISCUSSION

J.P. SEILER

Swiss Federal Research Station, CH 8820 Wädenswil (Switzerland)

The system testing for reverse mutations and using histidine-requiring mutants of *Salmonella typhimurium*, in short called the Ames test, is in widespread use throughout the world. It owes its success to its simplicity, rapidity, inexpensiveness and its clearly defined endpoint, as well as to its predictive power for chemical mutagens and carcinogens. With its common usage, however, problems of reproducibility began to appear, which were clearly shown in the results of the first European Collaborative Ames Test Study. It became obvious that certain fundamental microbiological aspects of the test system were only poorly understood or investigated and that thus, despite the detailed protocol given by Ames, willingly or not, modifications were introduced whose effects on the result were unknown. As stated in the Introduction, much of the variability of the results could be traced to the use of liver homogenate, whose biochemistry is even less understood and investigated. However, also in the area of microbiology, causes for such a variability could be found, and these have been discussed in the 6 foregoing contributions. Some of these are of a more theoretical kind, others of most practical importance. However, theory and praxis mix to varying extents in the areas discussed here. There are 3 prominent areas, interconnected with each other, which are important to the outcome, the evaluation and the interpretation of the Ames test.

A first area covers aspects of the microbial cell itself. The number of spontaneous mutants per plate as a baseline measure for estimating the mutagenic response of a certain bacterial strain to a chemical substance is obviously of great importance and can influence the interpretation of a test result. Also the intricate mechanisms of mutation induction, fixation and expression may be influenced by the chemical treatment of the bacteria. A doubling of the number of mutant colonies per plate therefore does not correspond to a doubling of the mutation rate, thus giving only a qualitative measure of mutagenic potency. At the same time, this means — as was pointed out in the discussion, too, — that the use of "mutation factors" for the comparison of different compounds is not a valid method, since this ratio is influenced to a great extent by external and internal factors. This holds the more true as the response expressed as number of revertant colonies per plate can be influenced to a major extent by the number of bacteria seeded on the plate and by their physiological status.

In a second area problems of the statistical evaluation must be discussed together with the problems of interpretation of a test result. It has been shown that under non-standardized conditions of bacterial growth and cell number certain fundamental statistical assumptions, like the Poisson distribution of spontaneous colony numbers, are clearly untenable; even in a single laboratory, where conditions would be more uniform, this assumption cannot be maintained. Also the different shapes of concentration—effect curves allow only the general use of non-parametric statistical methods to test for significance. Such findings lead again to the conclusion that to our present knowledge the Ames test can be regarded only as a qualitative measure of mutagenicity.

What follows from the 2 above-mentioned areas is the third group of problems which is associated with the standardization of the Ames test. In this context 2 seemingly divergent opinions have to be reconciled. On one hand strictest standardization of the protocol seems necessary in order to obtain uniform results. In such a standardized manner every single laboratory is certainly working; however, in order to reach agreement between laboratories, standardization must not only be confined to the single laboratory's standard practice. An inter-laboratory standardization is then obviously needed. This need is most apparent in the case of weak mutagens, which in some laboratories are detected, while others label them as negative. Opposed to this strict standardization requirement is the need for flexibility of the protocol. Some compounds might need specially adapted procedures, as pointed out in the foregoing contributions, which cannot be accommodated in a strictly standardized protocol. Both of these opinions have their merits and should be incorporated into a meaningful execution of the Ames test. The solution to this ambivalence can come only from the first 2 areas discussed above. Since a strict standardization is obviously needed for the repeatability of experimental results, any deviations from such a protocol should be completely understood regarding their influence on microbiological parameters in order to make the experiment truly reproducible.

From this point of view and from the data presented in the various contributions to this Panel Discussion the following points can be said to be important sources of variability, which should be controlled in the best possible way, and which should be reported in the most detailed manner possible in any publication on the Ames test.

Careful avoidance of any possibility of mutagen formation in the process of preparing cultures and plates helps to keep background mutation frequency as low as possible.

The number of cells plated has to be as high as possible and should be recorded by means appropriate to determining living cell counts.

Growth conditions used in the lab should be checked for their influence on growth rates of all tester strains. The use of late log/early stationary phase cells instead of "overnight cultures" is strongly advocated.

Due regard has to be given also to the chemical purity of test substances, e.g. presence of mutagenic contaminants, to the toxicity and solubility, limiting the testable concentration range, and to the possibility of differential toxicity to his^- bacteria and his^+ revertants.

The size and set-up of an experiment should be determined in accordance with the statistical method to be used for the evaluation of the test results.

Science consists of ever asking questions and of never taking anything for granted. In the Ames test we have been guilty of neglecting these basic principles of true science. Taking the Ames protocol for granted we did not ask why it called for an overnight culture of the bacteria, what were the reasons for using 10^8 bacteria per plate, how to obtain such numbers, which would be the best growth phase of bacteria to be used, or what were the reasons for obtaining the specified number of spontaneous mutants. However, such questions are being asked now and the experimental data we were allowed to present here have shown that we are beginning to appreciate and to understand the microbiology behind the Ames test. We do not present final answers and we cannot give an ultimate recipe for doing the test. What we can (and hopefully did) is to direct the attention to certain problem areas and to present ways and means for improving the test and its inter- and intra-laboratory reproducibility. The Ames test is a very valuable tool for the detection of mutagenic and carcinogenic compounds; improving the experimental side of the test will then certainly lead to an improvement in the interpretability of its results. If our Panel Discussion led to the introduction of such improvements in the laboratories performing the Ames test, it can be said to have served its purpose.

CONCLUDING REMARKS

CONCLUDING REMARKS

PER OFTEDAL

Institute of General Genetics, University of Oslo (Norway)

As some of you may remember, I made something like concluding remarks in Tucepi last year. I interpret the Program Committee's invitation to repeat the performance to mean either that I did well and so seemed suitable, or that I did poorly, but deserve another try.

Obviously, during these splendid 3—4 days of rather hard scientific work we have heard far too many talks and seen too many posters for discussing all papers now. There have been 118 posters presented, and nearly 100 oral presentations. On looking back over the annual meetings, there is a general trend which is getting stronger every year, namely that of integration and generalization. This is in contrast to an earlier phase of casuistic and alarming reports concerning single substances, or new test systems. But in spite of the enormous amount of new material brought to light, progress is still slow in specific areas. This is clear after hearing the excellent report on Captan by Dr. Legator in this meeting, and comparing with Bridges review from 1975. The uncertainties of a risk estimate for humans are more or less unchanged.

Many papers deserve to be mentioned and discussed. The overview given by John Ashby of the international collaborative study of paired carcinogen/non-carcinogen compounds gave new information even for those of us who have heard similar reports before. Ian Purchase discussed the data in Oslo earlier this summer. The specific and the generalized observations, which came out in this study, confirm and detail many of the advantages of the short-term tests, and also reveal some weaknesses. Most interesting in Ashby's presentation was in my opinion the conclusions he drew regarding test tactics: a broad battery of cheap and rapid in vitro tests directed towards suspect substances. If suspicion is confirmed, use a battery of short-term in vivo tests. Only with this data set in hand would it seem sensible to decide to perform or not perform a long-term cancer test, which may cost half a million pounds.

Morris' review of the effects of Benomyl on the mitotic apparatus gave — together with Kappas' paper on the genetic effects of Benomyl — an analysis of one of the presumably most important aspects of mutagenesis in man. And this is mutagenesis which is not caused by damage to DNA, but to proteins. The phenomenon of widely used fungicides conceivably causing damage of great clinical importance in man is indeed disturbing.

The symposium on food and gut content of mutagens might frighten one away from many pleasures. At the same time these findings highlight the problem of action against factors or contaminants that are integral parts of man's way of life. This problem is quite different from the equally difficult one of making a risk benefit evaluation of pesticides like Captan, amitrole, or nitrofen — of increasing carcinogenic potential.

Yet, regulatory action must come — with regard to be it hair dyes, vinyl chloride, or Benomyl — in one form or another. And to justify and codify such actions one must go through processes which on the one hand must produce a simple or modified yes/no response, on the other hand draws on a more complex and wide base of data and politics in addition to what may be called the scientific one, that we usually pride ourselves with being in charge of, and which ICPEMC surveys, as reported by Professors Sobels and Bridges earlier in this meeting.

Public concern and pressure is strong, and the need for regulatory action is great. To a large extent due to the new short-term tests, new principles of evaluation need to be established. Over the past 5 years some of us have been concerned with, and many of us have seen and discussed the various guidelines which have been proposed in several countries. One of the most significant documents is the OECD proposal on toxicological testing which will be finalized during the coming 6 months.

A few points concerning this document might be mentioned here. Genetic toxicology is only a small sector of the total toxicology. The base-line requirements is still only one point-mutation test and one chromosome abnormality test, though a number of other tests are mentioned in relation to more extensive testing. This is to be expected, since several of our people have been involved in the development of these documents. The bulk of the document is concerned with other toxicological tests, and I shall not discuss these, except to mention that the 3-generation rat test appears to have disappeared, which I think is good. It often carries a sort of genetic overtone, though of course it is singularly poorly suited for testing anything of the sort.

But, let me point out that this is an administrator's document, seeking to lump things into simple categories, for the purposes of handling, trade and disposal according to understandable and usable regulations, to protect society and the individual. In other words, complexity and detail are sacrificed for simplicity and efficiency. We are then in a sense back with John Ashby and his suggestions for a more rapid and efficient test tactics, possibly at the risk of loss of some precision.

To some, this may not be a tasty recipe. Obviously, the detailed analysis of a suspect mutagen, through many modifications and many systems, is an enjoyable exercise and an intellectual challenge. On the other hand, when the end-point is to be a legislative yes or no, one may wonder to what extent experimental refinements are justified. It should also be noted that the level of refinement required is already considerable. Metabolic pathways, organ specificities, species differences, etc. are all implicit in the required array of other tests. The choice can be seen as between a socially meaningful exercise, and a total scientific immersion.

Science has some very fundamental rules, one of the most important being that of approaching one's problems with an unbiased mind, and using unbiased instruments. Therefore, I was a little surprised and critical when reading in the OECD document that the results of the gross pathology findings — say at autopsy — should be communicated to the microscopist, so he would know what to look for, and where. The possibility for independent and unbiased judgement seems to be curtailed by this procedure.

But then I came to think of the extreme bias we all show — in one sense — when we expose our super-sensitive and super-sensitized test organism to suspected mutagens under conditions designed to give maximum effects, sometimes even in non-coded situations.

It does not follow from this that the results are not true or that they are less useful.

But it might be a subject for thought — on the plane going home, or in the Agora this afternoon — to try to decide where science ends and social responsibility takes over, and how wide is the area of overlap. I think I'll leave you with that thought.

AUTHOR INDEX

Andrae, U., 129

Ben-Gurion, R., 11

Carere, A., 87

Evans, H.J., 111

Garner, R.C., 33
Göggelman, W., 129, 173
Grafe, A., 167
Green, M.H.L., 159
Greim, H., 129

Hardell, L., 105
Hesse, S., 129
Hill, M.J., 41
Hryniewicz, M., 79

Kappas, A., 59

Loprieno, N. (preface), v

Matsushima, T., 49

Mattern, I.E., 187
Moreau, S., 25
Morpurgo, G., 87

Obe, G., 19
Oftedal, P., 197

Ramel, C., 69

Schwarz, L.R., 129
Seiler, J.P., 151, 155, 191
Sielenska, M., 79
Sugimura, T., 49
Summer, K.H., 129
Szymczyk, T., 79

Thompson, M.H., 41
Trojanowska, M., 79

Venitt, S., 3
Vollmar, J., 179

Zalejska, M., 79
Zdzienicka, M., 79

SUBJECT INDEX

Acetaldehyde, 19—22
    acetaldehyde hypothesis, 21—22
    carcinogenicity of, 21
    chromosomal aberrations induced by, 20—21
2-Acetylaminofluorene (2AAF), 138
Aflatoxin B1 (AFB$_1$), 33—38, 140
    carcinogenicity and mutagenicity of, 34, 35
    metabolism of, 34, 35
    reactions with nucleic acids of, 35
    synergism with hepatitis B virus of, 34
Aflatoxins, 33—38
    carcinogenicity of, 33, 34
Ames test, 4, 12, 50, 79, 88, 91, 94, 95, 97, 98, 101, 151—193
    culture conditions effect on, 173—178
    evaluation of, 187—189
    growth media effect on, 155—158
    number of bacteria effect on, 173—178
    spontaneous mutation effect on, 159—164
    statistical problems in, 179—186
    test cell problems in, 167—171
2-Aminoanthracene (2AA), 151, 176—177
2-Amino-3,4-dimethylimidazo[4,5-*f*]quinoline(MeIQ), 51—53
3-Amino-1,4-dimethyl-5*H*-pyrido[4,3-*b*]-indole(Trp-p-1), 49—53
2-Aminodipyrido[1,2-*a*:3′,2′-*d*]imidazole-(Glu-p-2), 51—52
2-Amino-6-methyldipyrido[1,2-*a*:3′,2′-*d*]-imidazole(Glu-p-1), 51—52
2-Amino-3-methylimidazo[4,5-*f*]quinoline-(IQ), 51—53
3-Amino-1-methyl-5*H*-pyrido[4,3-*b*]indole-(Trp-p-2), 49—53
Antioxidants, 11—17
    carcinogenesis inhibition by, 11
    colicin-inducing activity of, 15—16
    deactivation of mutagens by, 4, 11
    mutagenic activity of, 16
Aromatic amines, 121
Arylesterase, 71

Aryl hydrocarbon hydroxy base (AHH), 123
Ascorbic acid, 12, 14—15
    colicin induction by, 15
    DNA interaction with, 12
Aspergillus nidulans, 60—66, 73, 81—82, 88—89
    diploid strains of, 62, 63, 89
    haploid strains of, 66, 88
    parasexual cycle of, 61

Benomyl, 59—66, 90, 92—93
    chromosomal aberrations by, 65
    mutagenicity of, 65—66
    recombinogenicity of, 60—64
Benzimidazole fungicides, 59—60
Benzo[*a*]pyrene, 11, 43, 115, 120—121, 132
Botrytis cinerea, 65

Captan, 91—93
Carbedazine (*see also* MBC), 59—65
Chlorodinitrobenzene, 136
1-Chloro-2,4-dinitrobenzene (CDNB), 176—177
Chloroform, 139—140
Chlorophenols, 105—108
Chloroprene, 133—135
    mutagenicity—carcinogenicity of, 134
Cigarette smoke carcinogenicity, 111—127
    synergism of, 112
Cigarette smoke mutagenicity, 111—127
    chromosome damage, 117—127
    sister-chromatid exchanges, 116—126
    smokers urine mutagenicity, 115—117
    sperm abnormalities, 113—115
Colchicine, 64
Colicin-induction test, 12—14
    advantages of, 14
Colon cancer, 4, 5, 41, 42, 49, 107
Colorectal cancer, *see* Large-bowel cancer

Dichloroacetaldehyde, 71, 76
Dichloroethanol, 71
Dichlorvos (DDVP), 69—77, 91—94
    carcinogenicity of, 75

Dichlorvos (DDVP), (continued)
  demethylation of, 71
  metabolism of, 70
  mutagenicity of, 72—74
  phosphorylation and hydrolysis of, 70
3,4-Dicyclo-pentenopyrido[3,2-a]carbazole (Lys-p-1), 51—52
Dithiocarbamates, 79
DNA-damage repair test, 90
DNA repair, 132—133
Drosophila, 73—74

Eremofortins (A, B, C, D and E), 25—26
  see also *P. roqueforti* metabolites
Escherichia coli, 6, 7, 73, 161
Estrone, 131—132
  carcinogenicity of, 131
  inactivation of, 132
Ethanol, 19—22
  carcinogenicity of, 21
  mutagenicity of, 19—20
  synergism with smoking, 21
  teratogenicity of, 21—22
EUE cells, 97—98

Faecal bacteria, 41—46
  dehydrogenation by, 42
  dehydroxylation by, 42
Faecal bile acids, 41—46
Faecal extracts, 4, 9
Faecal mutagens, 3—10
  Ames test, 4
  antioxidants-dependent, 9
  chromosomal test, 9
  diet-dependent, 9
  chromosomal test, 9
  fluctuation test, 9
Fanconi's anemia, 20
Fetal alcohol syndrome (FAS), 21, 22
Fluctuation test, 7, 38, 65—66, 159—164
Fuberidazole, 59, 60
Fusarium oxysporum, 65

Griseofulvin, 65
Gut bacterial metabolism, 41—46

Halogenated ethylenes, 140
Herbicides, 91—102
  aminotriazole, 91—93
  atrazine, 100—101
  dalapon, 92
  diquat, 97
  paraquat, 97—98
  phenoxy acids, 105—108
  picloran, 91, 93
  tordon, 91
  trifluralin, 98—99

High doses of mutagen—carcinogens, 129—143
  effect of contaminants at, 143
  effects on metabolism of, 135—138
  extrapolation to low doses from, 130—131
  threshold doses, 131—132
Human cancers, 49, 54, 112
Hydroxyl dehydrogenase, 42
Hyperthermia hypothesis, 114

Inactivation of carcinogens, 131
Inductest, 80—81

Jonckheere test, 183—186

Kruskal—Wallis test, 183—184

Lacuna, 12—14
Large-bowel cancer, 3—10, 41—42, 44
Lung cancer, 111, 123—125

Malignant lymphomas, 105—108
Manu—Whitney—Wilcoxon test, 183—186
MBC (methyl-2-benzimidazole carbamate), 59—65
Metabolic activation, 138—140
  impaired, 138
  species differences in, 138
Metabolic inactivation, 140—141
Methyl methanesulphonate (MMS), 72
Microtubule protein, 64
Mitomycin C, 14
Mitotic crossing-over, 61—64, 82, 89—99
Mutations forward, 88

Naphthalene, 137—138
Neurospora crassa, 60—73
Nicotiana alata, 99—101
Non-disjunction, 61—64, 82, 89—99

Organophosphate insecticides, 91—94
Oxygen reactive species, 132—133

Penicillium roqueforti toxin (PRT), 25—32
  biological activity of, 27
  cheeses and, 31
  induction of DNA-protein cross links by, 29, 30
  interaction with DNA of, 28
  interaction with replication of, 27
  interaction with transcription of, 27
  interaction with translation of, 27
  metabolic alterations by, 27
  metabolites of (*see also* Cremofortins), 25—26
  metabolization of, 30

Plant metabolism test, 99—102
Poisson distribution, 179—183
Purpurogallin, 14—17
    mutagenicity and colicin-induction of, 16—17
Pyrogallol, 14—16
    mutagenicity and colicin induction of, 16
Pyrolysis carcinogens, 49—54
Pyrolysis mutagens, 49—54
    amino acids and protein in, 50—54
    biological activities of, 54
    chemical structure of, 50—53
    cooked food and, 50—54

Risk-benefit evaluation of chemicals, 69

S9 liver homogenate, 80, 91, 115—117, 120—121, 191
Saccharin, 69
Saccharomyces cerevisiae, 65, 73
Salmonella typhimurium, 3, 5, 6, 11, 50—51, 71, 73, 74—80, 86, 88, 91—98, 115—116, 121, 151—193

SCE (sister-chromatid exchanges), 74, 116—126
Sodium ascorbate, 11, 14—15
    colicin induction of, 15
Sodium azide, 156—157, 168, 176
Soft-tissue sarcomas, 105—108
Sperm abnormalities assay, 83
Streptomyces coelicolor, 88, 91—96
Styrene, 140—141

Thiabendazole (TBZ), 59—60
Thiophanate, 59—60
Thiourea, 141—142
Thiram, 79—85
    metabolic activation of, 83—85
Threshold doses of carcinogens, 131—132
Tobacco tars, 115, 117, 123
Treat and plate test, 161, 163
Trichlorophone, 75—76
Tubulin, 65

UV-sensitive mutants, 66

Vitamin C, 12

X-Rays, 113